Information Experience

SUNY series, Studies in Technical Communication
───────────
Miles A. Kimball, Derek G. Ross, and Hilary A. Sarat-St. Peter, editors

Information Experience

The Strategy and Tactics of Design Thinking

CRAIG BAEHR

SUNY PRESS

Published by State University of New York Press, Albany

EU GPSR Authorised Representative:
Logos Europe, 9 rue Nicolas Poussin, 17000, La Rochelle, France
contact@logoseurope.eu

For information, contact State University of New York Press, Albany, NY
www.sunypress.edu

Library of Congress Cataloging-in-Publication Data

Name: Baehr, Craig, author.
Title: Information experience : the strategy and tactics of design thinking /
    Craig Baehr.
Description: Albany : State University of New York Press, [2025]. | Series:
    SUNY series, studies in technical communication | Includes bibliographical
    references and index.
Identifiers: LCCN 2024049481 | ISBN 9798855802740 (hardcover : alk. paper) |
    ISBN 9798855802757 (ebook) | ISBN 9798855802733 (pbk. : alk. paper)
Subjects: LCSH: Information behavior. | Information resources. | User-centered
    system design. | Communication of technical information.
Classification: LCC ZA3075 .B28 2025 | DDC 025.5/24—dc23/eng/20241206
LC record available at https://lccn.loc.gov/2024049481

Contents

Illustrations

Figures

Table

Acknowledgments

The author would like to give a special thanks to the contributors to this book:

Susan Lang, PhD, professor of English and director of the Center for the Study and Teaching of Writing, Ohio State University, for writing the foreword to this book, which provides a comprehensive overview of the approach, chapters, core contributions, and wider implications from an experienced academic perspective.

Mike Fuller for use of three original watercolor prints and Liz Pohland for use of two original watercolor prints as illustrative examples of how user perception and cognition function in the interpretation of visual information.

Carbon Sitars and Kiere Shaffer for use of original graphic artwork to illustrate successful strategic branding throughout their product line, website, and marketing campaigns.

The Society for Technical Communication for use of screenshots of the Technical Communication Body of Knowledge (TCBOK) project to illustrate examples of successful interface features and page design.

Foreword

SUSAN M. LANG

Technical communicators create more than content; they create holistic information experiences. While many disciplines may integrate individual practices and principles of developing information products and environments, technical communication brings together the core competencies of developing all aspects of an information experience.

The above sentences, which appear on the closing pages of this book's final chapter, reinforce the impact of this text. It is, in and of itself, a technical information product, in that it provides a comprehensive discussion of how developers and users create these products with varying levels of success, beginning with an encompassing definition of *information experience*, along with its five core components—perception, cognition, design, environment, and branding. In this book, Craig Baehr has drawn from his work in both industry and the academy to develop a comprehensive examination of information experience with significant implications for the field of technical communication.

Core Contributions of the Text

The first chapter introduces readers to the core concepts underlying information experience and demonstrates how the interaction of these elements enables the creation of useful information products and environments. It also unpacks the theories underlying each component in pragmatic ways and provides clear examples of application of each. In doing so, it provides detailed explanations of related terms—such as intended experience (IE) and perceived experience (PE)—and considers how correlations between these factors result in more or less effective overall experiences for users.

Subsequent chapters focus on a single concept and associated theories. The second chapter examines perception and visual thinking, both of which influence larger cycles of user interpretation, especially as more information experiences have moved to digital and hybrid formats. These digital experiences employ more layers of information, which pose additional challenges to our perception of both content and medium of presentation. While user perception has clearly evolved and adapted to digital information products, such as websites and mobile apps, such evolution poses challenges to developers who wish to augment environments with new features or abilities. Our initial perceptive reactions combine with our analytical processes to understand information presented.

All information experiences are, to some extent, learning environments, as users encounter different organizational patterns, navigation tools, and content in their quest for information. Cognitive processes are examined in more detail in the third chapter, particularly exploring their impact on visual thinking. While perception represents the first level of encounter with an information product, cognition quickly kicks in to help users process and interpret these initial impressions. Baehr explains how cognitive processing as a multilayered, complex activity can lead users to interpret the same experience in a variety of ways, thus adding to developer challenges. Furthermore, our processes when dealing with a current situation call on past cognitive filters, preferred learning modalities, and learned experiences, all of which add to the differential reactions and impressions of a particular information experience.

While the prior two chapters focus on how users may interpret or respond to information, the fourth chapter turns to the context of the experience, the information environment. In many ways, this may be the most essential section of the text, given the evolution of information environments from print/analog to digital in the past 30 years. As Baehr notes, the environment functions as the interface between user and content, whether that interface is a digital or other artifact. Central to his discussion, though, is understanding how hypertext theory provides the underlying framework and rationale for many of the features users now expect from their information experiences, particularly in digital environments. Hypertext theory has emerged as one of, if not the, primary ideas underlying 21st-century technical communication. The fundamental hyperlink enables the creation of multilinear, multilayered, and multiauthored information structures that seamlessly incorporate video, audio, and alphanumeric components. Hyperlinking enables developers to create the possibility of highly customizable user experiences while providing a more authoritative route through the product, depending on each user's needs. Hypertext theory also demonstrates how these characteristics evolved from the conventions and constraints of earlier print-based information products. Baehr also discusses the convergence and evolution of communication technologies as he outlines typical properties of a hypertextual information environment and how developers customize and adapt those to suit their particular information product.

The fifth chapter focuses on information branding. While many may consider branding the purview of marketing and advertising, technical communicators have a role to play in developing the strategic branding and visual identity for the information product. Branding is all about establishing an impression in the user's mind. Baehr discusses how brands are strategically developed and tactically messaged to users via a product's visual identity, ideally balancing a consistent, yet creative and distinct message that emphasizes the product's value to users. Brands that work extend the initial perception and subsequent use of the information product and should reinforce initial perception of the product. If users receive contradictory or inconsistent information when encountering a product, either on a first visit or subsequent returns, their actual or perceived experience of the product and environment may affect their decision-making about whether to continue its use. Although branding should project consistency, the effectiveness of the brand should be revisited when new upgrades or iterations of the product are rolled out.

The sixth chapter turns to tactical design, which encompasses the disciplines of information design and user experience design. Information design is essentially the design principles and practices applied to the content of the product, whether it be textual, visual, or spatial. User experience design considers how the product will be interacted with by users for a variety of purposes. Those designing the content return to basic principles of perception and cognition as they strive to create products that are consistent yet visually distinctive in ways that clarify information to users. Their work is in part tested via the user experience design process, which focuses on the tasks that users will complete using the product. Successful information experiences result from products that are designed to be both usable and accessible to the greatest range of potential users. However essential, tactical design must be informed by strategic choices in creating information products that aren't simply used, but are also experienced holistically.

The seventh and final chapter emphasizes the holistic nature of the information experience, from developer intent to user perception and interpretation. Baehr reminds readers that successful information experiences are most likely to result from thorough and iterative user-focused research, attention to accepted principles of cognition and learning, and willingness to develop best-fit solutions that will be revisited and modified as needed, rather than ideal solutions out of the box. He concludes the chapter with several key takeaways: content is experienced, not simply read; information products and their experiences must evolve together; visual and spatial thinking dictates how we experience content; and finally, information experiences represent alignment between content, environment, and users.

As readers can imagine, these brief summaries provide a semblance of the text's coverage of each aspect of information experience. Throughout, Baehr demonstrates how to apply abstract concepts, such as hypertext theory and visual/spatial theories. To do so, he provides readers with abundant figures and screenshots to illustrate concepts and techniques. These serve as useful refreshers for those readers familiar with Gestalt and hypertext theory and as a clear introduction to readers who are newly exposed to them. While the book is not a textbook, the careful combination of visual and textual information enables it to function as a comprehensive overview of creating information experiences for technical information products. This alone makes the text valuable to a large chunk of its intended audience—current technical communication practitioners, especially those who need a concise way to articulate the core strategic and tactical elements used in developing technical communication products. Additionally, this text supports and aligns with each of the Society for Technical Communication's (STC) nine core competencies of professional certification, and can be explained as part of a constructed information experience, adding an overarching applied framework to the competencies. Such is significant on its own merits. But this volume contributes in other important ways to the field.

Wider Implications

Most significantly, this text elucidates the scope of what technical communicators are doing nearly a quarter into the 21st century. Having a better understanding of the breadth of technical communication roles and responsibilities has implications for trainers (both professional and in the academy) and future professionals, particularly given the academic/industry gap in the field in both publication and training.

While the publication gap has been well-documented from a variety of angles, it is ultimately less impactful to the field (in large part because the gap stays, well, a gap) than the discrepancies between the academic and industry perspectives regarding the education of new and continuing professionals. Moore and Earnshaw (2020) discuss inconsistencies in the ways that academic program directors, recent graduates of technical communication programs, and practitioners view the field in terms of definition, needed skills, and the value of various levels of professional development. Andersen and Hackos (2021) document ways in which academic research on technical communication often lacks relevance to and is inaccessible to practitioners, while Boettger and Friess (2020) find that a fairly small group of academic researchers drive much of technical communication's published research agenda toward areas (rhetoric, pedagogy) that are not served by empirical research, while topic areas more tied to practice (usability, editing and style, design, and knowledge management) are underrepresented in core journals.

This isn't a simple problem—the evolution of many current technical communication courses, majors, and minors at postsecondary institutions in the United States has occurred in particular ways to attempt to solve specific curricular and professional issues at those institutions. Two issues stand out—the need to provide students with writing courses, and the need to employ instructors to staff those courses. The uneasy resolution of both issues has resulted in undergraduate courses and programs that often provide students with writing experiences that will not reflect those they are likely to encounter in workplaces, in part because they hire instructors to staff these courses who are not, themselves, trained technical communicators. A secondary result is that courses or programs designed to teach students to write for a specific profession, say, engineering, may indeed do that—but these students are also not trained technical communicators. What does this mean in practical terms? And how does

Baehr's text provide an avenue for changing elements of these courses and programs? Regarding the first, larger issue—providing students with writing courses—we repeatedly observe the following:

- Departments charged with providing a first-year writing course have faced, over the last 30-plus years, a shrinking demand for that deliverable, with predictable budgetary outcomes. If these are universities, contingent faculty and graduate students face the loss of jobs; if smaller colleges, tenure-line faculty could be at risk.

- Parallel to this, public institutions face pressures to move students efficiently through degree programs, with some calling for combining writing instruction with a course currently part of the major. Yet faculty in many disciplines do not see themselves as writing instructors.

Often, the solution to this issue is that departments such as English or Communication develop individual courses in technical writing—a solution designed to provide students with writing opportunities and instructors or graduate students with jobs. The problem—most if not all of those assigned to teach those courses lack the breadth and formal training in actual technical communication theory and practice. Because of that, students who take these courses are subjected to whatever interpretation of the field drives syllabus development of a particular course or program. Even in most best-case scenarios, students may receive writing and editing practice in typical technical genres, but the other seven core competencies are barely treated. While this is not an unreasonable situation, since a single course only has so much time and bandwidth available, the issue becomes a significant problem when academic departments attempt to create a minor or major in technical communication without committing to a clear or holistic vision of what that means, and without faculty equipped to teach all nine competencies. If departments wish to provide comprehensive programs in technical communication, they need to recognize their disciplinary knowledge limitations and look to industry, as well as the academy, for assistance to develop a framework that informs these programs. Newmark and Bartolotta (2021) provide readers with one way of doing so, describing how they engaged with STC's core competencies to inform the learning outcomes for their foundation-level technical communication course, thereby creating a "through-line" for students from university to industry. This is a significant step forward, but Baehr's text enables us to consider even more.

I see this text as the first disciplinary framework of consequence proposed since Carolyn Rude's "Mapping the Research Questions in Technical Communication" from 2009. And it extends Rude's work in significant ways in that it provides organization at the theoretical and applied levels in considering how we view, teach, and research technical communication. Rude's overarching central question asked, "How do texts (print, digital, multimedia; visual, verbal) and related communication practices mediate knowledge, values, and action in a variety of social and professional contexts?" She examined books from the prior two decades to determine the directions researchers engaged to answer portions of this larger question and offered the following four lines of inquiry: disciplinary, pedagogy, practice, and social change. After exploring potential areas of research and offering research questions based on her findings, Rude reminds readers of two critical points. First, she notes that "interest in texts as they enable knowledge and action links academic and practitioner research" (206). Second, that "no one else pays such close attention to texts used to get work done, particularly work that requires specialized knowledge" (206), which she views as a unique contribution of technical communication. Baehr's taxonomy of information experience responds to both of these points in a generative, field-sustaining way in that he provides a foundation for further developing both academic and practitioner work in technical communication.

Finally, Baehr extends our previous work on hypertext theory as a foundational theory for technical communication. When we consider a few of the key characteristics of hypertext theory (collaborative authoring, content focus, hyperlinking, hypermedia, intertextuality, and multipathed structures), it is clear that many aspects of an information experience can be catalogued via the theory (Baehr & Lang, 2012, 2019). While many tend to equate hypertextual tools, such as HTML, CSS, and XML, with hypertext theory, it's critical to recognize that they are but a small part of instantiating a digital information experience. Understanding an information experience through hypertext theory, combined with theories of cognition, learning, perception, and others, enables developer teams to enrich user experiences and continue to improve technical information products for end users. It also provides academics a holistic framework from which to develop cohesive, interdisciplinary curricular programs, whether they be at the certificate, minor, or undergraduate or graduate degree level. Leveraging experts in perception, cognition, and hypertext theory, along with communication and organizational management experts, to create information experience curriculum that focuses on the unique properties of technical communication will ensure a bright future for the field. Baehr's text provides us with a blueprint for doing so.

Introduction

Successful experiences begin with both an idea and a user in mind. These experiences are, in part, constructed from how a user processes new information, contextualizes it, and forms concrete impressions. When users interact with an information product environment, they might have different motivations, purposes, or uses for it. However, *information isn't simply read or used; rather it creates a holistic experience for users.* Similarly, information products aren't simply for use, but they represent holistic experiences, which can create affect, immersion, utility, and wonder. As information products continue to require increasingly higher levels of engagement and interaction, our development practices in creating them must be understood from a user's perspective as much as it is from the intended product content, features, and outcomes.

Information experiences combine an understanding of users, an overall development strategy, and discrete tactics, which inform the process of designing information products. While a specific experience may be intended, often the user's interpretation may attribute different characteristics and impressions to that experience. From a user-centered design perspective, product development strategies must address customization, interaction, preference, and semantic components of an experience. This kind of user-centered research can be derived from an understanding of basic perceptual and cognitive processes, involving how sensory information is categorized, filtered, patterned, prioritized, sorted, and eventually, understood by users. The strategies, which evolve from our initial idea as well as this understanding, should incorporate methods of branding a product and how the product environment, or interface, provides useful functions, navigation, organizational patterns, and uses. Brands communicate consistent and coherent messages and themes, which are an extension of our original idea for a project. In short, they communicate a strategically idealized intent and an experience developers want for their product users. Similarly, environment design provides the landscape from which users interact with content and interpret basic function and format present in an information product, and must be designed with similar intent. At the level of implementation, strategies support the selection of specific design tactics, which incorporate various visual, spatial, and textual characteristics and features that enhance the experience. Holistically, the information experience is the result of these collective efforts and represents the most basic form of interaction between content, environment, and user.

This book provides a usable framework for creating and under-standing how information experiences are successfully designed, which can be applied to a wide range of information development projects and tasks, including application development, content creation, instructional use, and user experience design. Whether content is created by human, machine, or artificial intelligence, the creation of memorable information experiences begins with the user. The specific components of information experience extend from the user into the subjects of perception, cognition, environment design, strategic branding, and tactical design, which are discussed individually and through an integrative approach in this book. Collectively, these components form a framework for understanding how information products function on a level beyond utility, as a holistic experience for users.

Chapter 1

The Elements of Information Experience

We all experience information in different ways. While there may be similar perceived characteristics, features, messages, and themes commonly associated with a particular information product, our individual experience adds its own unique permutations. Information products and their environments, whether physical, hybrid, or virtual, also convey specific and sometimes different impressions to users, some intended, while others may not be. Our use of a product and its environment greatly affect our overall experience with it as well. Our uses with a particular product may align with our expectations or prior experiences, or we may discover new appropriated uses, which better suit our individual needs. Initially, we respond to new information products and environments in similar ways, based on our basic perceptual and cognitive processes, as well as our prior learned experiences. In turn, these prior experiences help us learn new information product features, which become even common or expected, allowing us to transfer knowledge from one experience to the next. We don't simply read new information, we experience it.

Our information experience represents the holistic combination of both user-driven impressions and developer-led features that characterize an information product and its environment. The information experience encompasses more than use, extending to our overall perceptual and cognitive impressions, which help us comprehend, construct, interpret, and learn from those experiences. Developers convey these experiences through established practices and processes, including information design and user experience design techniques, which help create highly usable information products for a wide range of users and technological uses.

Accordingly, the strategies and tactics used to design information represent holistic theories and practices of design thinking.

When we design and develop information products and services, we're creating more than technical content that is readable and usable, but rather, holistic information experiences that convey a broader range of characteristics, messages, themes, and uses. Technical communication involves creating and developing highly technical subject matter using a wide range of modalities, processes, and tools to create information products, which appeal to its specific users. These products serve a wide range of industries, including engineering, government, health care, science, information technology, and many others. While information product developers have specialized knowledge of the discipline and in the subject matter for an information product, technical communication skills also include techniques for developing and designing content that are adaptive—in ways that best fit the equally wide range of products and disciplines in which we work. These particular skill sets include technical expertise in information design, content authoring, information architecture, user experience design, content management, technical editing, instructional design, project management, web development, coding, and many others. As our methods and tools evolve, the related technologies and skills required and used in developing information products and environments will follow. Collectively, these changes help define both the disciplinary boundaries and unique skill sets of technical communicators, forming the basis of the techniques, products, and processes we use to create meaningful information experiences for our users.

These practices and processes are also informed by discrete theories and principles from an equally broad range of subject areas, including human perception, cognition, experiential learning, and hypertext theory, to name a few. In turn, our applied practices of these theories inform how we create, design, organize, and publish technical content that appeals to a diverse range of audiences and product environments. As both users and developers, how we perceive, think, and use a particular information product includes how we interpret the unique visual, spatial, and textual characteristics of both product and environment. When interacting with a new information product, our perceptual and cognitive processes help us form a conceptual whole that represents our information experience. And from this holistic understanding, we adapt and learn how to make information products useful to our individual needs. Our perceptual processes govern our visual focus, problem solving, identification, classification, and concept formation while our cognitive ones help us make meaning, comprehend patterns and structures, and make decisions in information environments (Baehr, 2019). We also create, filter, and sort experiences with information products through discrete learning behaviors, such as active experimentation, observation, and concept formation to learn from information products and environments (Kolb, 1976).

Our understanding of these processes can also be applied in many different ways, including in our design techniques (information and visual design), development tasks (user experience design), product branding (visual identity branding), and processes (iterative and agile). Information design practices include techniques for the successful creation of information components (documents, media, objects). These practices can also inform how information is structured (information architecture, organizational design, content management, content markup) and presented (visual and graphic design, scripts, style sheets, templates). For example, user experience design focuses on product usability, through an iterative process of researching, structuring, prototyping, and testing phases, which include the development of interactive components (content, navigation, interactions) for the product's intended users. In developing strategic brands for products, we develop characteristics, messages, and themes that inform how we develop an information product, or even an entire line of usable products. The processes we use to develop information products is also increasingly agile, following iterative development cycles and sprints that allow for flexible and process-mature workflows, information visualization techniques, prototype development, and quality assurance tests, which enable us to develop robust and engaging information product experiences.

When we interact with new information products, our perceptual and cognitive processes also help us think visually and spatially in understanding how they are designed, organized, and structured. Visual-spatial thinking helps us perceive properties such as depth, direction, and cohesiveness to understand the individual characteristics of both content and environment, such as how the various components relate and collectively function as a conceptual whole. Some of these characteristics we perceive include shape (object and size), position (axis and depth), contrast (color and shading), continuity (repetition and pattern), and many others. In electronic information environments, the visual and spatial characteristics often share symbiotic relationships, helping us understand how visual, spatial, and textual codes function collectively in basic concept formation (Johnson-Sheehan & Baehr, 2001). While our perception may recognize familiar visual-spatial characteristics, new and novel ones we encounter may change our impressions and subsequently, how we instinctively respond to and comprehend the overall information experience. Since visual-spatial thinking guides how users may instinctively respond to new experiences, these processes can also help developers create more meaningful information experiences and products, through practical application and techniques informed by visual-spatial theories.

Information Experience Core Components

Technical communicators and other information developers are not simply writing documentation for consumption; rather, they are creating holistic content experiences, which function beyond the level of use. Ultimately, the goal of technical communication is to create usable content and products; however, the information products they develop also communicate an information experience, which represents their impressions and interactions with a particular product. The combined characteristics, features, messages, and themes within these information products and their environments create a holistic sense of the information experience for the user. However, this information experience often changes throughout an information product's life cycle—as does its intended and perceived qualities, well beyond a single iteration. While applicable to other kinds of products and services, in this book, information experience primarily focuses on its relevance in information-based products and environments, whether physical, hybrid, or virtual. Ideally, any information experience begins deliberately and systematically, in the development process that is ideally aligned with user expectations, information needs, and various uses. This experience may be carefully planned and executed by product developers through various tasks, features, functions, and other aspects. These elements may also be extensions of specific, intended themes—such as adaptive, customized, intuitive, and responsive. But after a product is released to the public, users also form their own concept of the information experience, by interpreting the visual, spatial, and textual codes, through their own perceptual and cognitive processes, as well as collective experiences with other information products. Consequently, information products and their experiences are informed by a wide range of factors, which encompass the user, the environment, the content, and the product development strategies and processes. Specifically, an information experience encompasses five core components: perception, cognition, branding, environment, and design (see fig. 1.1).

Information experience includes how users interpret content, whether exploring, learning, or sense-making, through their perception and cognition. This understanding, along with product specifications, can support an overall strategy for designing information environments and a particular strategic brand, which is both product and user-centered. And the tactics of information experience involve the implementation of practices that support both information design and user experience design principles

Figure 1.1. The five core components of information experience. Information experiences are informed by theories of perception and cognition, environmental characteristics, strategic branding, and tactical design principles and techniques. *Source*: Created by the author.

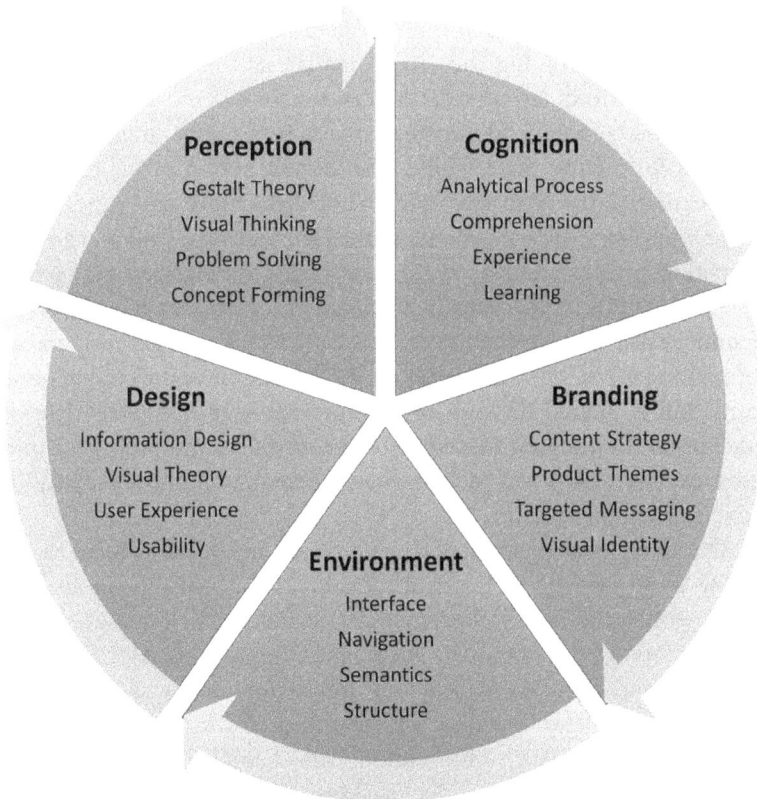

and standards. Together, these core components represent a cohesive process, from formulating an understanding of users, to developing an overarching strategy, and finally to implementing discrete tactics, which enable the creation of successful information products and experiences.

Holistic information experiences are shaped by both users and products, which can be understood through the various perception and cognitive acts, applied practices, processes, and theories that characterize user interaction with information. From the user's perspective, our visual and spatial thinking is an act, representing an iterative series of perceptual and cognitive processes, which subsequently influence our actions, behaviors, and impressions. From the product developer's perspective, this visual and spatial thinking translates into specific applied practices, which function as the design tactics used to create more user-centered information products and experiences, which align with their expectations and impressions.

The combination of processes for both perception and cognition describe how visual-spatial thinking operates. Our perception involves sensory acts of examining the visual, spatial, and textual characteristics and semantics within an information environment. Users sense, interpret, filter, sort, and comprehend information as a series of iterative acts, which help them form basic impressions. The Gestalt notion of the conceptual whole represents the collective outcome of these processes—understanding the categories, classifications, features, functions, and purposes of information present (Koffka, 1935). Our cognition engages our comprehension, analytical, and evaluative thought processes to match patterns and understand relationships as well as how we obtain, process, and use information (Barry, 1997; Richey et al., 2011). Cognitive processing of new information is filtered through prior experiences and preferred learning modalities and strategies. Understanding how users interpret these information experiences is just as critical as understanding how they use the products that convey them.

Our development practices are also greatly informed by these aspects, enabling us to develop tactics and techniques that help us design more user-centered products. Collectively, these tactics include several core principles used in the branding, information design, and user experience of information products, which help developers create characteristics, features, and messages that align with the ways in which users think, use, and ultimately experience information. For example, strategic branding includes the creation of specific messages and themes, which can be translated into specific characteristics and features to communicate specific product experience factors. Branded messages and themes, as well as information product environments themselves, possess their own unique characteristics and features, which contribute to the overall information experience. Information design principles and practices focus on specific use of visual, spatial, and textual codes and content, which communicate these experience factors in both the product and its environment. And user experience principles and practices focus on iterative prototyping and testing methods, which facilitate the access and use of products and their environments, with adaptive and responsive elements. Collectively, these tactics form a holistic triad, which, in turn, enables us to create successfully aligned information experiences for users.

Perceptual and Cognitive Processes

Perception and cognition form a symbiotic interchange of processes that govern our visual and spatial thinking when encountering new environments. Our information experiences with information products and environments are largely constructed from these processes, which help us interpret the various visual, spatial, textual, and other codes present. As such, our information experience with a product is often influenced by how we think when interpreting as well as using information products and environments. Visual and spatial thinking represents a set of behaviors and responses, which includes everything from our initial instincts all the way up to our learned behaviors and experiences. Users interpret content and environment, both separately, as individual elements, and collectively, as a conceptual whole. Perception is the first and initial component of visual-spatial thinking, where we interpret both new and familiar information in an information environment. As an active engagement of our collective senses, we selectively examine visual, spatial, and textual elements in our visual field, focus on specific objects, study figure and ground elements and their relational aspects, analyze shapes and objects altogether while attempting to grasp a holistic understanding or sense of the whole information environment. Rudolf Arnheim (1997) characterized these collective processes and acts as visual thinking, stating that visual thinking is as much about what we see as what we think. But in an electronic environment, such as a website, sometimes what we see can be augmented by features such as software tools, browsers, screen readers, and other tools and technologies—helping us effectively see more, or differently—enhancing our information experience with other layers of sensory input, such as spatial characteristics. In these environments, the visual and spatial become inextricably linked as we attempt to understand concepts, meaning, position, semantics, structure, and other elements in information products and environments (Johnson-Sheehan & Baehr, 2001).

Arnheim's principles of visual thinking also describe how human perception focuses heavily on visual codes and characteristics, yet they also suggest the importance of spatial codes and characteristics, which are interpreted alongside their visual counterparts (Arnheim, 1997; Johnson-Sheehan & Baehr, 2001). Visual elements tend to be noticed, selectively, based on their distinctiveness or novelty in a visual hierarchy of sorts—for example, we tend to notice animated, bright colored objects before static, dull colored objects, yet we may also notice their arrangement or position relative to others. We may focus, or fixate, on specific visual elements if we think they merit further study or use, yet sometimes interpret a group of elements as related based on their proximity. Our perception also involves discerning similarities and differences in both figure (foreground objects) and ground (background), yet we may also notice how they are grouped to suggested relatedness or disparateness, based on visual and spatial characteristics. We also look for familiar concepts in shapes and objects in our visual field to form meaning, yet their position, layering, dimensionality, or depth may also affect our perception and understanding of them. The culmination of our visual thinking is the formation of a conceptual whole, or holistic understanding of the information product and its environment. While perception may be the first, critical foundational step in our conceptual understanding of new information, what we perceive is also informed by our cognitive processes, prior experiences, and preferred learning modalities.

Our cognitive processes support visual and spatial thinking, working symbiotically with our perceived impressions, helping us learn new information. Our cognitive processes help us process patterns, characteristics, relationships, and changes in our environment so that we can better understand and learn new information spaces (Barry, 1997). These cognitive processes also support varying levels of complexity, from the memorization or conceptualization of basic knowledge or facts to higher level processes of analyzing and synthesizing information we encounter. For example, the act of pattern matching involves our ability to classify and identify objects and their unique configurations within those stored in memory, such as the arrangement of dots in a spatial grid. When we compare and contrast different objects, we notice similarities and differences between objects in the same space. We can filter and sort stimuli to help us identify, organize, prioritize, and discount individual items in an environment for further analysis. And we can build semantic categories, which involves a more complex sorting cognitive process, helping us organize information into conceptual groups based on their relational, semantic, and structural characteristics. As such, our perceptual and cognitive processes work together, and iteratively, to help us shape and refine our understanding of information environments.

Our cognitive processes are also influenced by prior learned experiences—what we've previously learned and committed to memory. Cognitive learning theory focuses on how we process and understand new information, and how it extends to how we use it (Smith & Ragan, 2005; Richey et al., 2011). In a basic sense, our cognition helps interpret what we perceive and translates it into learned experience. In turn, this experiential learning allows us to build patterns of behavior and reaction, which can be applied to use in actual working contexts. Kolb (1976) describes our experiential learning process as a cycle, which includes active experimenting, concrete experiencing, reflective observation, and abstract conceptualization—which underscores the importance of how we learn and process new experiences. Our learned experiences function as filters through which we can process new information and experiences. For example, if we can pattern match something to a high enough degree of certainty, we can act based on previous assumptions, or if we encounter significant differences, we can create new patterns to guide our behaviors, reactions, and use. Similarly, our visual-spatial thinking functions as an iterative exchange of perceptual and cognitive processes, which repeat continuously back and forth until we are satisfied with a conceptual whole that represents what we've learned. This thinking can also inform how we apply practices in developing information products with specific messages, design features, and contexts of use. Strategic branding, which includes the characteristics, messages, and themes created for a specific product, can also benefit from these applied thinking practices, enabling us to create more effective information products and messages. Developers can integrate these practices into the presentation and interactive properties of both content and environment, which are attuned to the ways in which humans think visually and spatially. Specific practices might include methods used to create, design, edit, style, and publish technical content, whether the deliverable is a document, website, video tutorial, procedure, presentation, or other visual, spatial, or textual form. Product development strategies might also include implementing these techniques into the larger, unified process of researching, wireframing, prototyping, and testing new information product for optimal usability and eventual use. Information products and experiences are informed by both strategic (developer or organizational paradigms) and tactical (appropriated techniques) factors (Kimball, 2017). As such, a wide range of disciplinary practices, including how we brand, design, and develop information products can be used to optimize the information experiences for their users.

How Information Environments Communicate Experience

Product environments are designed and developed by both strategies and tactics in creating successful information experiences for users. The product environment includes important characteristics, content elements, features, and tools, which enable users to access, explore, interact, and use information. An information product environment can be physical, hybrid, and even virtual, including a wide range of characteristics familiar to us from both print and electronic environments. These environments might include printed documents in various forms, but more likely electronic portable documents, such as PDFs, websites, applications, software interfaces, interactive media, videos, and other kinds of virtual environmental characteristics and features. Users also perceive visual and spatial elements differently in these various environments, which also changes our understanding of how they think about and, subsequently, use these product environments. In addition to the specific characteristics, features, messages, and themes present, it is important that we understand the capabilities of the product environment and how best to integrate content with environment in our development strategies. Users may want their products to be just as accessible, comprehensive, and useful in different product versions or iterations; however, there may be differences in how users perceive and interpret these environments as holistic information experiences. Based on similar experiences with other products, users may have different expectations based on the type of product environment, whether it is a mobile application, electronic document, software program, or website. Product characteristics and themes, such as being adaptive, flexible, navigable, and responsive—may differ in their implementation strategies and tactics, which are dependent on the product environment. For example, a printed version of a technical paper might be accurate, comprehensive, and informative, while its electronic version may also need to be navigable, responsive, and searchable, in addition to these characteristics, to satisfy both user expectations and product environment capabilities.

As such, product environments often require different thinking and implementation techniques in both strategy and tactics. The conventions and techniques used to develop a geospatial knowledge database, which can deliver multidimensional content experiences in real time, will differ from those used to create a print-based atlas, which might use the same data sources. As one example of an interactive content portal, the Internet Archive (https://archive.org) is an online, open-source web-based media archive, which users search and navigate through a collection of historical versions of different document and media types. Their collection includes a wide range of content types, such as electronic books, images, journals, movies, podcasts, software libraries, videos, websites, and other forms. Through this collection, users can view different versions of electronic documents and media, including how they have evolved over their life cycle as information products. The Archive's primary searching tool, the Way Back Machine, allows users to experience the collection through independent exploration and searching, which covers an extensive volume of archived content on a wide range of subjects. Another similar content collection, the Society for Technical Communication's Technical Communication Body of Knowledge (TCBOK) portal, functions as an interactive knowledge base, which functions as a real-time historical collection of electronic content and media related to the knowledge, practice, and theories used in technical communication (see fig. 1.2). The TCBOK's mission is to provide a body of knowledge for the field of technical communication, through its collection of articles, links, media, and other references, which can be used as a tool for learning, research, and professional practice (TCBOK, n.d.).

While these and other collections may vary widely in their content assets and features, their interactive nature as research databases require vastly flexible information environments for users to browse, search, and use material in these collections. Within many electronic (or online) clearinghouses of content, users can borrow books, download files, listen to music, browse websites, and perform many other useful tasks in their own independent research, whether academic, personal, or professional

Figure 1.2. The Technical Communication Body of Knowledge website functions as an interactive knowledge base of articles, links, and references that serves both academic and professional technical communicators. *Source*: "Web Page Design," Technical Communication Body of Knowledge, https://www.tcbok.org/development/web-page-design/. Used with permission.

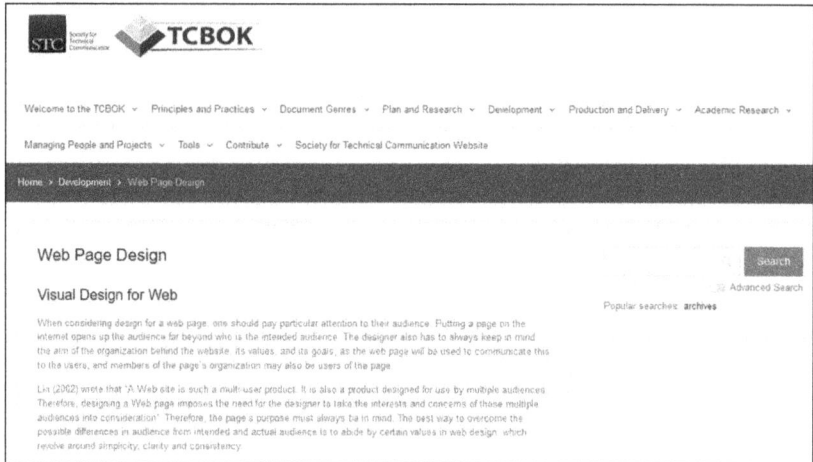

in nature. One unique and essential feature of these kinds of information environments is the use of hyperlinks, which convey semantic relationships between linked content elements, often in a nonlinear progression. Hyperlinks allow users to navigate different related content streams based on their keywords or relevance. While you could browse a stack of books in a library on different topics and publication dates in similar fashion, an electronic library may provide more flexibility in semantically browsing and searching through content, unrestricted by the physical limitations of a bookshelf or hardbound books. The information experience of such an electronic archive is vastly different from the physical library, which may include advanced searching options and other tools, which allow content to be explored rapidly, semantically, spatially, and even temporally. As with any web-based knowledge base or wiki, these content libraries often evolve rapidly over time by adding, categorizing, linking, and updating relevant information and resources, which are instantly available for use within the collection.

While information environments support performance and use of a particular product, such as a knowledge base, website, or wiki, the holistic information experience also communicated through the various visual, spatial, and textual codes present in both content and environment. Often, these characteristics and features change and evolve over the life cycle of a particular information product through subsequent iterations and updates. As the information product changes, the information experience will undoubtedly also undergo change. As an example, the Apple iPhone was originally conceived as having an intuitive interface, which most users could learn to use quickly through trial-and-error and basic recognition of symbols and keywords from other computing environments. The early Apple iPhone interface environment relied heavily on symbology and textual-visual pairs used for each of its apps, separated by borders and spacing to suggest different functions. This was vastly different from largely text-based menus and rudimentary iconography in previous cellular phone technologies. Another difference in the iPhone environment was the capability of opening multiple apps simultaneously, which allowed users to rapidly swipe and switch between content sources. These and other newer characteristics were implemented into later iterations of its design, which quickly revolutionized the market (and environment) for mobile phones. Over time, many of these novel features became conventional or commonplace, while newer ones were conceived and introduced. Other spin-off products, such as the Apple iPad tablet and Apple Watch, came later, which were created with a slightly different set of characteristics and features, while building on many of the core concepts of the original Apple iPhone product environment. With each new product environment and iteration, the information experience evolves with it. While an information experience may initially include a set of features perceived to be intuitive, adaptive, personalized, organized, and conceptual, ultimately these aspects continued to change in terms of their implementation in the product environment.

While hybrid and electronic environments rely heavily on information-based characteristics, often the physical characteristics influence the overall information experience for users. In the previous example, the Apple iPhone incorporates both physical and electronic-based characteristics, which influence its perceived uses. While the software interface is the primary means of communication and use, the physical characteristics incorporate the use of buttons, ports, and switches that augment and facilitate both access and use. These technological hybrids are not uncommon to the information product landscape, where content and environment can embody a combination of physical, hybrid, and virtual characteristics, which contribute to a holistic information product and experience. Other native electronic information products, such as web-based documents, digital images, and interactive media may also incorporate physical properties, such as display screens, keyboards, and mice as peripherals that support the information environment. Regardless of the information product, we must work within the capabilities and limitations of the environment to help with our appropriation and use of these products and their environments. As such, the perceived physicality of an information product environment is as influential as the hybrid and virtual aspects in shaping our information experiences.

How Strategic Branding Supports Experience

Information products are more than the sum of their features; they communicate successful and targeted characteristics, messages, and themes that contribute to the overall information experience for users. These product experiences are often conceptualized and planned through careful planning processes, including developing a holistic strategic brand that ensures these elements align well in a singular product. Strategic branding involves developing these specific messages, which communicate successful information product experiences. Strategic branding functions as a conceptual whole of sorts, encompassing the broad range of characteristics and themes used to develop an information product. Strategic brands, like information experiences, often evolve with product iterations, as well. Those changes may be based on several factors, such as how users appropriate products or how technologies used in their development change over time. Bloomstein (2021) suggests that while brand evolution is necessary, and even inevitable, some degree of familiarity must be maintained over time to retain trust and loyalty with users. Throughout this evolution, the information experience will still represent what is memorable to the user—including their impressions of the product, regardless of how favorable or unfavorable previous impressions might have been. While many messages and impressions are intentionally communicated through successful branding, others may be unintentionally communicated. While users desire a favorable experience, almost all information products and their experiences will have both strengths and weaknesses, some known and others not. While a well-branded product may create problematic information experiences, users may continue to be loyal to the product, creating acceptable appropriated workarounds for themselves, which account for these limitations.

In the software industry, information experiences are typically conceived as product experiences, which are developed initially through a strategic branding process. Strategic brands incorporate specific characteristics, messages, and themes, which inform the tactics of product design and development, including style sheets, templates, and visual identities. For example, an organization may want a specific product brand to be described as responsive, comprehensive, intuitive, modern, and adaptive. These descriptive characteristics and themes can also be informed or derived from other user research activities, such as existing product-line characteristics, previous product iterations, and other methods devised by the development team. A common technique for generating strategic branding themes is using product experience card sorting, from which developers select desired keywords that best represent the intended outcomes of the product (see fig. 1.3). In turn, these features can inform the development of specific product functions or messages, which can be communicated both explicitly and implicitly to the intended users. For example, to communicate responsiveness as a theme, loading times for pages and search results could be optimized for rapid access, based on a

Figure 1.3. Sample branded themes for an information product. Strategic brands incorporate various themes that inform the development of characteristics, features, and product messaging. *Source*: Created by the author.

specific metric or unit of time. A modern theme could be communicated through various design choices that include minimalist page layouts, neutral color schemes, or other supporting visual images that support that theme. Ideally, strategically branded themes extend beyond a single product into future iterations, throughout a product's life cycle, acquiring new characteristics and leaving others behind in the process.

Branded themes serve as guiding principles for the overall information experience like a roadmap, informing the development of specific features, objectives, and tasks and objectives, which align with these themes in very practical and tactical ways. For example, the navigable theme might be applied in developing searching and browsing tools for a website, perhaps in a site map that illustrates the visual and spatial relationships of content sections and pages, with embedded hyperlinks for ease of access. Similarly, the intuitiveness theme might be successfully implemented using shapes, icons, directional arrows, or performance tips that are familiar to users and contribute to ease of use with an information product. These examples also integrate visual and spatial techniques, which match both product development themes as well as users' visual and spatial thinking processes. Consequently, aligning strategic branding themes, which represent the developer's intentions, with typical user actions and thinking processes is essentially a user-centered practice. Developers can also use other supporting techniques grounded in design theories informed by the ways in which users perceive, understand, and learn new environments and products that support or align successfully with their strategic brands.

How Tactical Design Contributes to Experience

The successful design of information experiences incorporates an understanding of both user and environment, and applies specific strategies and tactics that help us create useful information products. Tactical design integrates established practices within specific contexts, which help us make decisions about how to implement our knowledge of both user and environment. These tactics include specific information design practices, such as the style, placement, depth, perspective, organization, graphics, and other aspects of content design. In turn, these techniques are informed by specific design theories and principles, which govern the visual, spatial, and textual codes throughout design work (Kostelnick & Roberts, 1998). While design principles are grounded in theory, their specific application or implementation may vary depending on product, user, environment, context, and other factors. Some examples of commonly used information design principles include the use of contrast, repetition, alignment, and proximity (Williams, 2015), as well as the use of alignment, balance, grouping, consistency, and contrast (Johnson-Sheehan, 2024). Although different design texts may propose different terminology and sets of specific principles, they have important theoretical foundations in Gestalt theory. Gestalt principles, supported by both perceptual and cognitive theories, include the use of proximity, continuation, completion, similarity, closure, symmetry, and figure/ground (Koffka, 1935; Kohler, 1947).

Regardless of the specific terminology or technique used, information design inextricably links visual, spatial, and textual codes within an information product in various combinations that apply to both content and presentation. While text may communicate a deliberate message in its use of diction, syntax, grammar, and other linguistic codes, our perception and interpretation of that message is augmented and shaped by other supporting visual and spatial characteristics present (as well as the medium, or environment present). For example, web development content markup denotes where and how headings, paragraphs, images, anchors, divisions, tables, and other elements are positioned and presented. While content markup alone may lack the desired design sophistication, they can be transformed when applying style sheets and templates developed using other scripting languages. These transformations enable both static and dynamic properties, which govern the color, shading, shape, position, depth, contrast, repetition, and interactive aspects to enhance the textual message. Whether visual, spatial, or textual in nature, these codes are all forms of content that work together to create holistic messages and information experiences for users.

Another applied design practice, user experience design, is also based on similar core features, which specifically focus on product usability and usefulness. The primary goal of user experience design is the creation of an iteratively prototyped and developed information product optimal for both users and context of use. The process of user experience design is also an iterative one, which requires continual evaluation and testing throughout the product development cycle, focusing on quality improvement and usability. User experience design typically includes activities such as journey mapping, prototyping, and iterative development and testing, all of which are ideally informed by user preferences and expectations throughout the entire development process. Some specific tasks might include the development of site maps (overall structure), task flows (procedural or process-oriented activity paths), annotated wireframes (interface layouts), and incremental prototyping (working models) that support product development (Unger & Chandler, 2012). Developing structure for an information product might include designing its information architecture, which maps out how content and processes are organized into logical pathways (Morville & Rosenfeld, 2006). Designing task flows or progressions can help developers map out how processes are ideally completed by users within the information environment, step by step. Developing wireframes involves creating basic interface layouts of an information product, such as how the actual information environment may appear in early form (Morville & Rosenfeld, 2006). And prototyping includes the development of actual working models of information products, which can be used for evaluation and testing of specific features or functions to maximize usability. With its focus on users, experience design can be informed by how users think visually and spatially in complex and interactive information environments, which can optimize use.

Collectively, these practices of information product design encompass how we develop both content and environment of information products, which are communicated through various visual, spatial, and textual codes and messages. Information design incorporates core principles and practices that focus on conceptual, consistent, positional, relational, and visual distinctive features of both presentation and content. User experience design follows an iterative process and set of principles that emphasize accessible, findable, responsive, and universal features in creating a usable information product. Together, these applied disciplines and principles can help developers apply practices informed by established design theories, as well as how users think and use information products, to create the best possible information experience for users.

When Information Experiences Go Wrong

Our impressions of an information product and its environment are shaped by more than simple use, encompassing our understanding of the content, design elements, features, messages, organizational techniques, themes, and other aspects. The information experience encompasses a holistic impression, whether we conceive it to be comprehensive, helpful, intuitive, modern, organized, or even useful. Sometimes, information experiences can vary quite far from the characteristics, messages, and themes that were intended. Our experiences can be strongly positive or negative, depending on how we perceive and interpret our interactions with information products and environments. Sometimes these impressions are created by unforeseen or unexpected elements, or in some cases, by intended characteristics or features of products that were designed to help rather than frustrate users. Regardless of intent, an unpleasant information experience can create a lasting impression just as strongly as a positive one.

For example, consider the case of a long-time e-commerce website that sells both new and used books, which has a well-established customer base. Users visit the site frequently to search for the latest and hard to find rare books and the searchable database the site provides has been described as accurate, easy to use, and rapid in its response to users. The website also has a well-established brand that uses a bookshelf logo, familiar slogan, and neutral color palette that is easily recognizable and present in all its marketing messages, including emails, product packaging, and website content. From user feedback gathered, the information experience suggests that the site and its contents are comprehensive, efficient, helpful, organized, and highly usable. For the most part, this information experience has been consistently maintained and has helped build the site's brand and reputation as a reliable online bookseller.

Recently, the development team conducted some benchmarking research on a few competing bookselling websites, noting that a few of them have redesigned their site, including the use of new images, updated logo, and reorganized navigation menus, as well as color, font, and other stylistic changes. As a result, several new members of the development team have been suggesting the website could use a fresher design and look, as well as changes to the navigation tools used to search and browse the site. The team also reviewed the latest results of a user feedback experience survey and after making a few recent changes, identified some minor findability problems with the navigation toolbar, where some users had difficulty locating links to specific content pages. In general, there were no major concerns about the overall design and presentation of content in the site, so the development team felt confident in their recommendations to update the site design.

As a result of the research results, subsequent meetings, and suggestions provided, the development team decided to proceed in refreshing the design of the site and update the navigation toolbar menu as the primary goals for the site update. From these goals, the team developed a list of feature changes and decided to update the bookshelf image used in the logo, replace the bland neutral color scheme with a lively, high contrast warm color palette, update the font faces for the major site headings and slogans used in the site, and reorganize and optimize the listing of links in the site navigation toolbar. The team developed a few prototypes of redesigned pages and tested them internally, among the team, for consistency and usability. They also decided to reorganize and rename some of the links used in the main navigation toolbar they felt would improve browsing and searching for users. Confident in their ability to continue to produce a well-designed and well-organized site, the team quickly created new site assets and developed a marketing campaign announcing upcoming changes and improvements. Several email messages and website news notices that were distributed had already generated some initial excitement from user feedback received on the site's message board.

Weeks later, the redesigned site was launched, and the development team was confident in their finished product. Despite the efforts of the team in the redesign project, shortly thereafter, negative user feedback and error reports started to steadily increase in the first two weeks of use. Users indicated they had trouble finding links to content pages because they had been moved to a different location or removed completely. Web analytic reports from user searches indicated an uptick in access times and errors, as well. From a design standpoint, users were mostly satisfied with the changes to the color scheme and other design elements and indicated only minor dissatisfaction with the logo and font changes. Some users questioned the need to change the site design and organization, commenting the site had always been consistent and reliable in the past. While generally, the design changes conveyed a new and refreshed appearance, they quickly became associated with the frustrations and findability problems users were reporting with the reorganized navigation toolbar. As a result, the information experience changed significantly for users based on mostly mixed reviews. Users characterized the new site as comprehensive, freshly designed, findability challenged, and slightly frustrating to use. Despite the intentions of the development team to improve the site, they had misappropriated some of the benchmarking and user research in making decisions for the redesign project and had relied more on their own judgment for some of the changes in the design and navigation of the site.

These problems are not uncommon and typically indicate misaligned product messaging, user research integration, or even misinterpreted messages or expectations. Many of these issues also demonstrate how problematic development practices, product messaging strategies, and mismatch with user expectations can create parallel impressions of the overall information experience for the intended users. In some cases, the information product's content may be the sole focus of development, irrespective of its impacts on users or in communicating a holistic information experience. While every information product may not be a perfect balance of user-centered and developer-centered preferences, it is important to consider how to clearly communicate messages through information products and their environments, which align in ways that users may perceive and understand them. Equally, this underscores the importance of understanding how information experiences are fully conceptualized and how to apply specific practices and techniques that are informed by users, environmental features, and product specifications.

Perceived and Intended Information Experience Factors

Ideally, information experiences incorporate clearly communicated features and messages, which help users easily adapt and use information products for a wide range of purposes. While developers may have the intention of aligning information products and their contents with user preferences, this may not always be successful in actual implementation. Despite their best efforts, some messages may be misinterpreted or missed entirely by users. Regardless of the range of characteristics, features, message, and themes intended, ultimately, the user's own interpretation can affect how an information product is received. As such, the information experience includes elements that are both known and unknown, as well as intended and unintended. While developers may have deliberate messages to communicate, they may not always be on-target, particularly when considering how users may interpret those messages. As a result, a user may miss critical information, such as steps in a process, or useful hyperlinks that help them search and browse information. For example, intended features may include accurate and timely content updates, as well as less desirable ones, such as frequent changes to navigation tools and design layouts, which may frustrate users. To accommodate these limitations, users may develop their own workarounds, initially unknown to developers, which help mitigate unintended frustration and improve their own performance needs. Sometimes, unintended but potentially offensive language or imagery could create problems, particularly with users from different cultures, where messages and symbols can have different meanings. Or, some features may also be known by developers because they were intended or discovered through iterative user feedback; however, there could be characteristics that remain unknown to both user and developer, which later emerge over a product's lifespan and use. While this suggests a certain measure of volatility between intended and known aspects, it underscores the importance of an iterative development process that includes rigorous user research and testing to ensure a more positive information experience. When creating information products, it is important to consider how information experiences are both intended and perceived, including how they contribute successfully to those products. Developers, brands, and the products they represent are crafted with specific characteristics, features, messages, and themes; however, users form their own impressions from these elements.

The intended experience (IE) is characterized by messages that are successfully communicated to users. Intended experience elements are typically part of the overall strategic brand, including slogans, themes, messages, functions, and other content. For example, a wiki-based information reference may include specific messages, such as a list of content topics to browse, and implicit ones, in the form of characteristics, such as accuracy, responsiveness, and ease of use. These characteristics can be recognized through actual use or observation, such as interaction with specific functions, navigation tools, or other content. An intended experience might be a result of careful planning or even serendipitous luck. Developers may also have enough creative flexibility within the scope of a project to successfully implement elements that successfully communicate the brand to users with or without considering their feedback or preferences. Intended experiences represent messages that are both intended and known to the developer, but may or may not be recognized or known to the user.

The perceived experience (PE) evolves exclusively from the user's perspective, which includes how messages were interpreted and whether or not those messages were actually intended by product developers. Perceived experiences can also include unsuspected miscommunicated messages, either unintended or unknown to the developer. For example, a software tutorial with a help feature that repeatedly returns inaccurate keyword search results on troubleshooting tips may not be intended by the developer, but rather the result of problems with the search function. Users interpret experience from use and assess product value based on intended features as well as unintended messages or problems. Regardless of awareness, intent, or visibility on the developer's part, these issues affect the user's perceived information experience of the product. Perceived experiences represent messages that are known to the user, but may be intentional, unintentional, or unknown to the developer.

As a result of the differences, the combination of what is known and unknown, as well as what is perceived and intended, comprise an information experience. Not unlike like the psychological concept of Luft and Ingham's Johari window (1961), which focuses on similar known and unknown characteristics of the individual's personality, information experiences have different components that are known and unknown to humans, whether they play the role of developer or user. Information experience, however, is about information products, rather than individuals, and how those experiences are communicated to the intended users. The characteristics of an information product that are perceived as intended demonstrate good alignment, where both developer and user share equal knowledge about the messages communicated. When the intended features of a product are more dominant and less informed or reliant on its users, the experience may be somewhat more system-focused, where the developer has a clearer understanding of features than the actual user. The characteristics perceived that exceed the boundaries the messages intended create a more augmented experience, where the user attributes characteristics and impressions to a particular product that the developer may not expect. The features that remain undiscovered by both developer and user, contribute to the unknown experience, which may be uncovered by later use, testing, or other method of discovery. All information products likely have communicative features and messages that undoubtedly fall into each of these categories in different quantities, which may be considered to be aligned, system-focused, augmented, and even unknown to both user and developer. Ideally, information products that iterate and incorporate user feedback throughout their development cycle can help create more strongly aligned experience factors, thereby minimizing elements that may be misaligned or unknown.

IE + PE: Aligned Experience Factors

Information features and messages that have a high degree of correlation between the intended and perceived experience demonstrate good alignment between developer, product, and user. Ideally, these well-aligned factors are the desired outcome where information products communicate clear messaging and can be used as intended. When developer intent and user perception are closely aligned, these factors contribute to positive information experiences for users. Well-aligned factors often result in products that demonstrate high levels of findability, intuitiveness, reception, and even use. For example, when a developer intends an experience to be accurate, intuitive, organized, and useful, and a user perceives those same qualities through interaction and use, the information experience is ideal for both parties. Well-aligned products may also demonstrate high success rates, favorable ratings, and possibly encourage helpful constructive feedback from users. However, in some cases, the messages communicated may not always be ideal in terms of their perceived value, although accurately conveyed. Some messages may have been intended and perceived but create unintended or unfavorable impressions. As an example, a developer may include performance tips that are written with humorous quips intended to entertain, which cause annoyance or anger with some users. Particularly, if this humor is used as part of important performance information, it may potentially offend users who misinterpret context from such messages due to differences in primary language or cultural background. Therefore, aligning intent and perception often requires both careful planning and user research to be successful in overall intent. Alignment should always consider the success of both message and its reception.

IE > PE: System-Focused Experience Factors

When information products have successfully communicated intended messages and features, normally this is ideal. But sometimes, those messages may dominate and even circumvent user information preferences or needs. As such, the perceived experience may be interpreted as more system-centered or system-focused, rather than user-centered or user-focused. System-centered approacheos place the developer, rather than the user, at the center of the development and design of an information product (R. R. Johnson, 1998). Sometimes a system-centered approach is necessary to help users learn a new system or feature, but experienced users may expect something more tailored to their needs and uses. System-focused features of a product experience may also have successfully conveyed messaging, but may similarly lack a user focus that tailors the product more toward their expectations or preferences. As a result, the developer's concerns dominate the product development, potentially resulting in a diminished information experience for users. For example, developers may organize a navigation toolbar in a website in ways that make perfect sense to the design team; however, this approach results in problems for users that have difficulty using the toolbar to navigate the site. The further disparity between the intended and perceived characteristics may result in user dissatisfaction and frustration. As a potential workaround, system-focused characteristics and features present in an information product might be enhanced with elements that support user expectation or performance over time, to help better align these experience factors for users.

PE > IE: AUGMENTED EXPERIENCE FACTORS

Sometimes our perception creates impressions that differ from the intended messaging of an information experience. In these instances, the user may interpret characteristics and messages that were unintentionally conveyed by an information product. As a result, product features may create impressions or unintended surprises for both users and developers. These augmented experience factors change the intended experience into one that may vary widely from what a developer may have intended. For example, when users perceive a new or revised product feature as problematic, such as how to block unwanted text messages on their phone, they may seek out their own processes or methods to achieve the intended result. Users may create their own creative workarounds, such as installing other apps or disabling messages from unknown users. While such workarounds may be successful, they may also cause users to ignore features or miss other available solutions. These augmented experience factors may also require remediation if they create performance issues or cause users to miss critical steps or content in an information product. When such a disparity between the intended and perceived messaging is realized, it is important to learn how and why these differences exist. Comprehensive and periodic collection of user feedback and product testing may help developers identify these augmented experience factors and determine solutions that demonstrate better alignment between product and user. Although augmented experience factors may be unexpected and less desirable to both developer and user, they can provide opportunities, or discoveries, for developers to learn more about how users appropriate information products in ways not considered. Augmented factors may contribute to improved future product features or iterations, which respond better to the needs of users.

IE ≠ PE: Unknown Experience Factors

Information products may also have hidden, latent, or even undiscovered messages by both developer and user. These unknown experience factors may remain obscured until they are discovered through use, feedback, or future iterations of a product life cycle. Undoubtedly, most information product will have some unknown factors, which have unknown influences on the information experience. For example, an information knowledge base may have content topics that contain broken external hyperlinks, which were updated by external website hosts, which are no longer accessible in the knowledge base. These broken links may be the result of modified or deleted content, or carried over from a previous iteration when they were functional. These unknown elements may be eventually discovered through repeated product use and testing, by either the developer or user at any time. Similarly, sometimes mature products experience feature creep, where an information product or system adds new features without a useful purpose, which are ignored by users. Developers may not realize these features are rarely used unless discovered through analytics research or by collecting user feedback. In turn, users may be oblivious of these features, unless they encounter them during use. Some unknown features may even be buried deep within the coding level of an information product, created by previous developers, and simply forgotten. As a result, unknown experience factors can create unknown performance issues or false impressions of an information experience until they are resolved. However, once discovered, they can provide potential opportunities for discovery and improvement of a product experience. One way to minimize potential unknown elements within an information product is to perform regular and rigorous testing and to collect regular user feedback. While these activities may not mitigate every possible unknown, it may help to discover some unknown factors so they can be addressed in future product iterations to enhance the information experience.

Holistically, an information experience may be characterized as a combination of these factors, with specific features and messages that are well-aligned, system-focused, augmented, and unknown (see fig. 1.4). Information experiences can be influenced by each of these factors independently, and in different ways, yet in most user-centered design practices, the primary goal is to successfully align product and user as best as possible. While this may be achieved through a variety of planning and development tasks, there is also some value in discovery through

Figure 1.4. Information experience factors. Information product experiences are a function of both intended and perceived factors. *Source*: Created by the author.

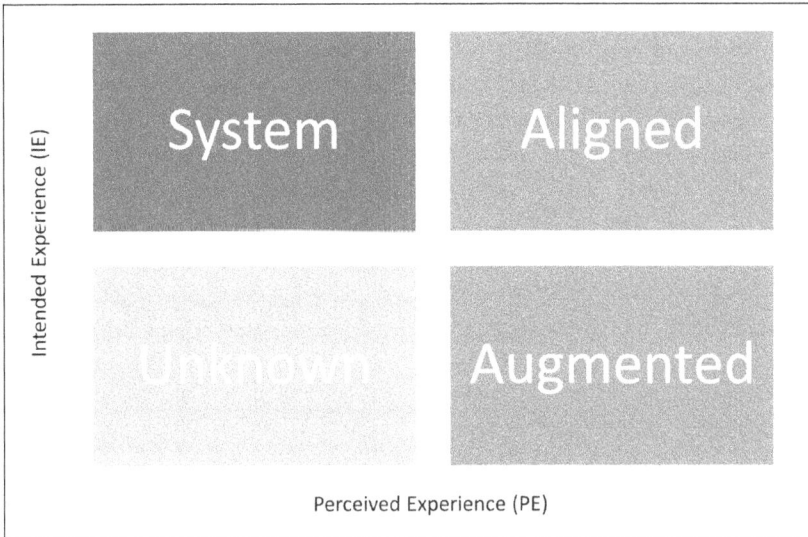

collecting regular user feedback and iterative product testing. In some cases, certain features or messages that are system-focused, augmented, or even unknown may have an intentional purpose. For example, in a website, certain content may need to remain hidden (or unknown) to users until certain conditions apply. Online message boards, which allow users to view and post comments for discussion may require a valid user account login or payment to access (or reveal) those features.

Online training courses may restrict access to content, such as advanced material, which can only be accessed once completing prerequisite modules. Augmented experience factors, perceived initially only by users, can have value once discovered by developers; they can provide valuable feedback on how to innovate and improve an information product. When users find a new workaround to complete a process, it can be included as a new feature in a future product iteration, once developers discover it through feedback and testing. System-focused factors can also be useful, particularly when introducing new features or functions to users. Once the user learns these new system functions, they can optimize their subsequent interactions with the product.

While many factors contribute to communicating an information experience, it may simply be impossible to account for every permutation or use. Information products also have their own relative importance and value independent of users, which follow the priorities laid out by developers and organizations that create them. Sometimes information experiences have characteristics that are trade-offs, which might cause unfavorable impressions, but satisfy a particular requirement, such as a regulation or standard. As with any product, a best-fit approach, which satisfies these conditions, should be carefully planned and executed to create successful information experiences and products. Developers should also consider each of these product experience factors throughout the product life cycle, so they can create optimal messaging and information experiences for users.

Holistic Information Experiences

Information experiences encompass user, environment, and the design of content. These experiences are informed by what we know about user perception and cognition and the information environment. In turn, this knowledge informs our applied practices of strategic branding, information design, and user experience design when creating information products. Users conceptualize product experiences and environments through their interpretation of communicated codes and messages. Visual-spatial thinking describes the combination of perceptual and cognitive processes users rely on to interpret information products and environments.

Perceptual processes describe how we process and prioritize visual information, solve problems in both simple and complex environments, understand relationships between figure and ground elements, and form concepts from the information we encounter (Arnheim, 1997). Together, these acts help us conceptualize an information product as a complete whole, representing our holistic understanding of the environment and the experience we perceive. Despite the actual form or modality of a product, whether physical, hybrid or virtual, our perceptual processes adapt to different environments and characteristics we encounter in the information landscape. For example, electronic (or digital) information products, such as websites, interactive media, and augmented reality environments, have unique semantic, relational, visual, and spatial properties that affect how we perceive and interpret them. Whether we're browsing or searching a website, moving back and forth along a timeline in an interactive video, or examining geopositional objects and data on a virtual street map, our perception of each of these environments differs, altering how we think and experience those information products. As we become more experienced with these and other content forms, our perceptual acts will continue to adapt and evolve as we learn from newer forms of content. Chapter 2 explores the theoretical foundations of our perceptual processes and how they inform our information experiences with products.

Cognitive processes describe how we make meaning from what we perceive, including acts such as pattern formation, concept recognition, experiential learning, and holistic understanding. These processes govern how we filter, sort, understand, learn, and ultimately behave (or act) in information environments. Cognition helps us make meaning from visual, spatial, and textual codes we perceive, in an iterative exchange of sensing and sense-making. The exchange between our perceptual and cognitive processes are virtually instantaneous, helping us conceptualize information experiences. In more complex information environments, such as web-sites and virtual spaces, our cognitive processes help us understand more complex semantic and structural characteristics, such as how information is layered, linked, organized, and interrelated. Cognition also functions at different levels, from simple knowledge acquisition to complex analytical and evaluative ones (Bloom et al., 1956). Cognitive processing might also differ based on the level of complexity, interest, or need users have in different information environments. Our individualized experiences and learning processes also play an important part in our sense making of information, acting as filters through which we process information. As a function of our cognitive processes, our experiential learning involves how we form concrete experiences, actively experiment, form concepts, and reflectively observe in new information environments we interact with, use, and appropriate (Kolb, 1976). As we make meaning and learn from our environment, what we learn can also be unlearned, relearned, or augmented, based on new conditions or interactions with an information product or environment. Chapter 3 discusses the important theories and concepts that dictate how our cognitive and learning processes contribute to our understanding of information experience.

Information experiences are conceptualized by users, through these processes, but they are also crafted by the developers of those information products. Information experiences are, in part, deliberately planned and developed, by both strategic branding and the information environment itself. Strategic brands are built upon specific characteristics, messages, and themes that communicate information experiences. These brands are informed by a wide range of factors including user research, product benchmarking, established standards, and actual product use. Specifically, understanding how users process information can also help brand development by selecting messages and themes that facilitate both use and understanding. Branding strategies can ideally inform all aspects of the product development process, including content creation, environment design, prototyping, and visual design. While overall brand messaging is strategic, an important tactical component is an information product's visual identity. Visual identities incorporate the use of various visual, spatial, and textual codes and styles, which successfully communicate branded messages (Baehr, 2007). Successful and sustainable brands also rely on a balance of both consistency and creativity throughout their development and scope (Airey, 2019). Over time, these brands may convey consistent messages or themes, which become recognizable or expected, among users. While users appreciate consistency, they often expect or appreciate fresh, creative changes over time. Sustainable brands often rely on creative innovations to keep information products up-to-date and aligned with the evolving expectations of users. Therefore, successful information product brands require proper alignment between user, product, and branded messages. Chapter 4 explores characteristics and techniques of creating strategic brands that form the basis of successful information experiences for users.

The information environment, or interface, is a way of seeing and experiencing the whole, connecting user and content in a holistic, seamless experience (S. Johnson, 1997). Environments can be physical, hybrid, virtual, simple, or even complex in nature, requiring users to learn the unique and different interface characteristics in information products. Information product interfaces that are electronic (or digital) in nature often incorporate content forms and features including the use of hyperlinks, linear and nonlinear structures, associative cross-linking, customization or personalization, interactivity, complex navigation, collaborative structured authoring, multimodality, and component content (Baehr & Lang, 2019). The specific implementation of these interface characteristics and features may vary, depending on factors such as user research, project scope, resources available, technological capabilities and limitations, and others. Over the life cycle of an information product, the interface environment may also have convergent features, which build upon previous iterations to support user performance in various contexts (Jenkins, 2006). Information products also integrate features built upon conventions and expectations, which are familiar to both developer and user to optimize the information experience. Creating information environments that support user performance is also essential in maximizing both the usefulness and usability of information products for users. Chapter 5 discusses how information environments and their unique features contribute to our understanding of the information experience.

Our understanding of user, brand, and environment will also inform our design and development practices, helping create information products that communicate a successfully aligned information experience. Specifically, information design and user experience design are two such development methods, based upon well-established theories and practices in technical communication, which help developers create highly accessible and usable information products. Both information design and user experience design focus on the design, functionality, and usefulness of both content and environment, ideally with the user in mind. Information design focuses on developing the visual, spatial, and textual characteristics of information products, but also the conceptual, consistent, positional, relational, and visually distinct properties of content. Within information design, specific design principles provide the broad theoretical guidelines, from which specific conventions and techniques are implemented into product design. Design conventions often take the form of style sheets and templates, which support the overall visual identity, or design concept, for an information product. Consequently, visual identities are a holistic representation of the information design of a product, which helps communicate the various characteristics, features, and messages. User experience design focuses on how specific content elements function and support actual use in the information product environment. As a user-centered design practice, user experience design emphasizes the importance of user research, wireframing (or structuring), prototyping, and testing, as iteratively repeated tasks that help create highly usable information product. The core principles of user experience design emphasize characteristics and techniques that help make products and their environments more accessible, findable, responsive, and universally intuitive in nature. While different approaches exist in how these tasks are executed, user experience design focuses on the primary goal of producing a product that synchronizes elements optimally to influence the perception and use of such products (Unger & Chandler, 2012). This synchronization underscores the importance of integrating practices that align user, environment, and content. Chapter 6 discusses the theoretical foundations, core concepts, and applied practices of both information design and user experience design, and how they support the overall information experience for users.

As a collective integration of user, environment, and content, the information experience evolves through the interpretive processes and applied practices of both information product users and developers. Our understanding of perception and cognition can help inform our development practices to create information products that best align with how users interpret and interact with product environments. In turn, this understanding can help both users and developers with similar products and features they encounter, develop, and experience. Whether we're using techniques of information design to position, style, or brand content, or carefully synchronizing elements through specific user-centered design practices to maximize product usability and usefulness, we're creating information products that aren't simply read by users, but rather, are experienced holistically. And through this deeper understanding of how information experience operates, we can create information products that better serve the needs and interests of users.

Chapter 2

Perception and Interpretive Experience

Perception has been studied by a wide range of fields from the biological to the technological and from art theory to information design theory. Perception, to an extent, characterizes our thinking, both visually and spatially, when interpreting new information. Our perception processes and filters incoming stimuli that help us make sense and refine our understanding of the world around us. In information products, these stimuli include primarily what we see, but also what we hear, and touch, and could also conceivably include other senses such as smell and taste, depending on the products and environments in which we experience them. While our perception is initially governed by instinctive responses, over time we can adapt our interpretations and actions to learn from other information experiences. However, our perception is just one part of a larger cycle of interpretive processes, including cognition and learning, which forms our understanding of an information experience.

Our information experience is also shaped through our unique appropriation and interpretation of information products we encounter. Perceptual processes govern how we make sense of the various visual, spatial, and textual codes present in an information environment, whether we're viewing a page, screen, or simulacrum. An understanding of how these important processes function can also inform our design and development methods of information products, including which design principles and conventions we use to create them. Whether these products are physical, hybrid, or virtual in nature, perception helps build our understanding of information product environments and their diverse interface features and modalities.

While visual perception is the primary focus of perceptual studies, what we see is often influenced by other sensory inputs. Other sensory stimuli can be part of the information product itself, or they can be part of the environment in which we experience content. For example, our online shopping experience may be primarily focused on the content and presentation of various products and services, yet we are also acutely aware of our surroundings, including the physical devices used to interact with the site, such as a mouse, keyboard, or touch screen. This experience may be further influenced by the smells and tastes of what we might be eating or drinking while shopping. Consequently, these additional stimuli can augment our information experience and behavior to an extent, influencing what we decided to purchase, study, or even ignore. Additionally, other technological aspects can influence this experience, such as assistive technologies, customized settings, browser plugins, or other such tools that help users *see* different layers of information within a single space. These tools function primarily to augment or enhance what we see to improve our product experience, but sometimes they may create frustrations or problems for us. For example, the online shopping site interface may have navigation tools and layouts which are overly complex, cluttered, or organized in ways that may confuse, complicate, or miscommunicate an information experience. When we encounter such problems, we may look for procedural workarounds to help solve problems, but after subsequent perceptual challenges, we may even seek out new online sites that better serve our needs and expectations.

Our perceptual processes affect our information experiences and are explained by two significant theories, which including Gestalt theory and visual thinking. Gestalt theory suggests our perceptual processes are highly deliberate, structured, and adaptive in how interpret the meaning of visual and spatial characteristics, such as shape, form, position, distinctiveness, and occurrence (Koffka, 1935; Kohler 1947). Visual thinking, applies these concepts further, suggesting our perceptual processes is a form of thought, which includes how we analyze, process, and comprehend simple and complex information through its visual, spatial, and textual codes (Arnheim, 1997). Our visual thinking is also adaptive and evolves through subsequent exploration, cognition, and learning of an information environment (Baehr, 2007). Particularly, in hybrid or electronic environments, the visual and spatial are inextricably linked (Johnson-Sheehan & Baehr, 2001). In these perceptual processes—our interpretation of the visual codes (color, font, shading, contrast) are also influenced by spatial codes (position, dimension, depth, proximity), as well as textual codes (words, phrases, headings, descriptors). In these environments, content can be adaptive, animated, customized, interactive, multidimensional, multimodal, and even responsive to specific conditions or user actions. These differences underscore how our perceptual thinking, and overall information experience, might differ from conventional printed or static information products. For example, websites organize content using one or more organizational patterns (hypertextually, hierarchically, linearly, and in custom configurations), through which we navigate, search, sort, and explore content through the use of various menus and interactive tools (Baehr, 2007). Information wikis, websites, and online databases typically integrate more hypertextual or nonlinear methods of interaction, linking, presenting, and structuring content. In these information environments, we may discover new possibilities, or serendipities, not previously conceived of within more conventional information products.

As such, our perception of these environments represents a fundamental shift in how we interpret information experiences from one media form or modality to others. We see many examples of this in everyday practice, where encyclopedias have been replaced by wikis and product catalogs have been replaced by interactive websites. Ong (2013) argued that many of these differences also are reflected in the authoring, development, and presentation of content as a natural evolutionary shift from one dominant communicative form to another (oral to written, written to print, print to electronic). Furthermore, Bolter (2001) describes this shift in how web-based information products as remediation, where newer content forms augment (and borrow) characteristics from older ones throughout their natural development cycles. Throughout the life cycle of an information product, some characteristics may expire, while newer ones emerge. This shift also occurs in how users perceive these newer content forms as information experiences.

While Bolter's concept of remediation suggests a continual evolution of information products, it can also be technological shifts that affect the changes. As technologies evolve, information product features will change as a result, as will, ultimately, our perception of them as both products and information experiences. For example, the information experience of using a telephone has evolved significantly in a century. Telephones were introduced as mechanical devices with buttons, dials, and wires with no visual display. As technologies evolved, they became portable (or cordless) handheld units with simple digital displays, which evolved into full-color screens with rudimentary display capabilities. Through further technological changes, they evolved into mini personal computer systems with high resolution displays, cameras, interactive maps, and many other useful features. Telephones have eliminated the need to memorize phone numbers, addresses, and the like. They also function as complex research tools, cameras, video editors, navigation systems, and communication devices, which can feature multiple modalities and content forms. The shift into the electronic age transformed a largely mechanical device into one that integrated many of the technologies that followed. As a result, this transformation also changed the way we perceive the information experience, as well as how we appropriate and use the telephone. As such, our information experience evolves with information products from their initial creation, through each new version and iteration. While we may perceive and interpret characteristics independently, we also see its collective features as a conceptual whole. Through our perceptual (and cognitive) processes, we come to understand the collective characteristics, features, and codes (visual, spatial, and textual) as a holistic information experience.

This holistic information experience is constructed and supported by our sensory perception in many important ways. Our perception helps us analyze, filter, interpret, and sense, which in turn guides our responses. We collect information from our sensory inputs, process that information internally, and output appropriate actions or responses. According to Kohler (1947), perception is inherently a Gestalt process, whereby we attempt to formulate a holistic sense of information presented through these various processes. Our perceptual processes also continually refine and restructure our understanding of information over time. Regardless of the medium, format, features, unique characteristics, or changes to a piece of information, our perception helps us adapt to changes in the information environment. Knowledge of these perceptual acts and principal functions can help developers better understand their users and how to align information experiences with ways in which users perceive them. This chapter discusses the importance of perception in information experience, specifically how principles and Gestalt theory, visual thinking, and perceptual processes inform our understanding of users, content, and environment as they relate to information products. It also explores how perception functions as visual and spatial thinking, helping users construct a holistic understanding of those products as meaningful information experiences.

Understanding Visual Perception

The ultimate goal of our perceptual processes is to form a conceptual whole or complete understanding from the individual and collective character-istics present in an information environment. In this sense, perception is more than seeing (or sensing); it is also an act of intelligent thinking. Barry (1997) also suggests the notion that perception is a problem-solving process, which she characterizes as visual intelligence. This suggests that our perception is both a seeing and a thinking process, which is in essence visual and spatial thinking, whereby we sense, interpret, and form a con-crete, holistic understanding of information environments we encounter. Our perceptual processes rely on two major acts—to collect sensory information and to interpret that information. The first part involves our basic perceptual responses to stimuli—what to focus on, study, discount, or even ignore—while the second part involves comprehension. These two acts are iterative exchanges that suggest perception is both adaptive and evolutionary. While we have standard reactions to stimuli encountered, we can change how we respond to certain stimuli as we learn from new experiences and stimuli. Perception is also analytical and holistic. Our perceptual processes work closely with our cognitive processes iteratively.

For example, when examining a painting or other work of art, our senses may notice differences in background details, colors, figures, shapes, and other characteristics both abstract and concrete. Through and iterative exchange of both perceptual and cognitive processes, we may form concepts from shapes or symbols observed, classify the painting styles or techniques, study relationships between elements in the foreground and background, compare and contrast various objects to others we've experienced, and other such acts. Collectively, these processes work toward a holistic under-standing of the work of art through examination of its characteristics and features. As we continue to study other angles or perspectives, our processing helps refine our understanding of the painting and its unique features. Some of our impressions will be stored in memory, while others may be discounted. Consequently, our cognitive memory functions as a conceptual library of sorts, working with our perception to recall specific features or impressions for subsequent analyses, as we continue to pattern match, understand semantics, and formulate new impressions.

Within any visual composition, the characteristics, features, and techniques used, such as the use of negative space or a gradient fill background, support the formation of a distinct, perceived impression or concrete whole, such as a shark floating in ocean waters (see fig. 2.1). The shark figure and ocean background are depicted using different color and shading techniques, which add both depth and realism to the image. The gentle gradation of color used in the ocean background may suggest differences in depth, lighting, or even positional perspective. The use of textured white splashes and strokes around the perimeter of the image suggest continuation of the image beyond its physical borders. And the use of darker color shading toward the bottom center would appear to

Figure 2.1. This image of a shark floating in water incorporates different artistic and compositional techniques, which influence our perception of depth, environment, perspective, and other features. *Source*: Mike Fuller, *Shark* [Acrylic print], 2023, Society6, http://society6.com. Used with permission.

depict a sunken ship, which adds complexity to the image. We might perceive the image not just as a shark in water, but perhaps setting a particular mood or tone, such as one that is adventurous, dangerous, or potentially fearsome. We may form other impressions based on how the image is framed, positioned, or displayed on a wall or in a gallery. Even our current mood or emotional responses may be imprinted on our overall perceived experience. Therefore, our perception is both adaptive and evolving and our impressions often are an influencing factor in our information experience.

As an adaptive process, perception helps form our innate reactions to new stimuli; however, we can override these basic perceptual responses, based on our experiences over time with similar stimuli (Baehr, 2007). We may change how we react to certain stimuli based on specific conditions, contexts, or situations for a particular case, while our initial perceptual reactions remain constant for any new stimuli. But based on certain recognizable characteristics, we might change how we perceive or process new information that bears similarity to prior experiences. For example, when confronted with minor functional usability problems in a software program, such as broken navigation hyperlinks or faulty search results, our perception can look for other tools or methods that help create workarounds for these problems. While specific hyperlinks or search results might not provide the expected results, you might try using other hyperlinks or key-word search terms, or even other tools to assist in finding more accurate results, using advanced search tools, subject indices, menus, or even try clicking on a few links from the search results to explore other options that might be available. We adapt by developing new or alternative strategies to search for information and solve problems we may encounter. These perceptual workarounds, of sorts, enable us to change how we react to these problems and look for new possible solutions or conceptual wholes. When encountering a problematic information product environment, such as a software program or website, our perception, as an adaptive process, contributes to and refines our overall information experience.

Our perceptions also have the potential to be altered every time we encounter information, which suggests our perception is also evo-lutionary. As our perceptual processes adapt to new information, they may evolve as we observe new characteristics, features, messages, and patterns, which may cause us to behave or react differently in the future. For example, advertising banners placed on web pages often obscure our view of information, particularly when they are positioned in noticeable locations. If we perceive these ads as irrelevant, we can override our natural perceptual response to notice them, and instead ignore them. We may develop workarounds to block, close, or obscure these ads from our view in subsequent interactions. However, if we perceive these same ads to be relevant at a later time, we may change our response to them as needed. Accordingly, perception helps determine the usefulness or relevance of visual stimuli, continually evolving based on future interactions to help improve our information experience.

Together, the perceptual acts of sensing and interpreting work together to help us form a conceptual whole (or gestalt) of information environments. Our perception helps us form visual hierarchies and prioritizes our focus area as we examine specific visual, spatial, and textual characteristics in an information environment, such as shape, form, color, style, and position. But perception can also help us functionally, to identify tools, functions, concepts, and relationships in these environments to solve problems or specific information needs. Perception also helps us interpret spatial characteristics, which include positional, relational, organizational, and even, temporal aspects of content. This helps us perceive dimensionality in information we see whether the environment is physical, hybrid, or virtual. For example, while much of our information experience is virtual in searching and browsing information in a website, there are specific physical aspects of which we may or may not be acutely aware. Conversely, the virtual aspects of the site may suggest three dimensional spaces, despite the flatness of a display screen, through the use of multiple web browser tabs, layers of information, pop-up windows, interactive graphics, and navigation tools. Our actions may also suggest perceived spatial qualities, as we swipe, stack, click, and mouse over elements on the screen. Our keyword search results may appear as a list of relevant search results, but their use of embedded hyperlinks allow us to move spatially from one source of information to another instantaneously, in a virtual simulacrum of content. Furthermore, these results might include temporal metadata, such as the specific time and date, which allows users to navigate multiple versions of content evolved over an information product's life cycle. Perception is an essential process helping us form a conceptual gestalt from the various visual, spatial, and textual characteristics of information, presented in a wide variety of configurations. It also supports our visual and spatial thinking in information environments, which is critical to our interpretation of information experiences.

Gestalt Theory and Human Perception

Our initial perception of stimuli begins physiologically, through our sensory inputs—visual, auditory, tactile, olfactory, and taste, with the ultimate goal of understanding of the whole effect of our environment. Within information environments, our perception may focus more prominently on specific sensory information, such as visual or auditory stimuli present. However, this information may be augmented or even influenced by other sensory inputs around us. Initially, we may have natural instinctive responses that guide our perception, such as what our eyes focus on, fixate on, how we perceive depth and shape, and how we understand the whole—which are all inherently perceptual processes (Arnheim, 1997). But our instincts and actions adapt and evolve over time to help us comprehend and learn from any changes in the perceived environment. Kohler (1947) suggests that our perception actively seeks a fully realized conceptual whole of an environment, information or otherwise, in addition to understanding its individual features. Our perception may involve studying the characteristics of color, distance, shape, spaces, and styles that are present, but ultimately, these collective features are pieces of a larger perceptual puzzle that form a conceptual whole, or gestalt. This gestalt represents our complete, holistic understanding of the various elements present in any environment.

Gestalt theory was developed in the early 1900s, based on the research on human perception by Kurt Koffka, Wolfgang Kohler, and Max Wertheimer. As an offshoot of psychological studies, Gestalt theory applied the theories and concepts on human perception, which influenced other fields including art history, psychology, and design, and resonated throughout the latter part of the twentieth century. Gestalt theory posits that specific acts of human perception work collectively to help us form conceptual wholes, that enable us make sense of the world around us (Baehr, 2010). The principles of Gestalt theory describe how perceptual acts function as both physiological and psychological processes, which include how humans classify, organize, sort, and conceptualize information to form complete, holistic understandings of environments. Gestalt theory serves as the foundation of many information design principles, heuristics, and applied practices, used widely in instructional texts used in technical communication today. This prominence suggests the importance that theories of perception, and in particular Gestalt theory, have in both our application and understanding of information design, and by extension, information experience. Principles of Gestalt theory include concepts such as closure, continuation, figure/ground, proximity, and similarity. These principles function independently and interdependently, helping explain how we classify, conceptualize, organize, and understand information environments and experiences. Understanding how these principles contribute to human perception and the formation of conceptual wholes can be also useful in our design practices.

The principle of closure refers to our ability to see a portion of an object, which may be partially obscured or incomplete, and perceive the object as its natural whole (Koffka, 1935). For example, we may see a car parked on a street, but in our field of vision, a large tree might be blocking part of it—yet we still perceive the car as a whole object, rather than just a part of it. Koffka (1935) suggests closed objects are perceived as more stable elements and naturally preferred over unclosed ones. Closure can also be affected by our perspective—certain viewing angles may obscure an object, affecting how we perceive it, and, ultimately, whether we perceive it as concrete (something recognizable) or abstract (something unknown). Closure is an example of how our perception helps us form conceptual (or stable) wholes from partial or uncomplete objects, functioning as perceptual shortcuts, of sorts. As another example, when staring out at a landscape with the moon behind a cloud or the horizon, we form closure of the object, perceiving it as a full, rounded moon, although a portion of it cannot be seen. Similarly, in an information product environment, magazines often use overlapping elements on their covers, such as text and images, which allows our perception to understand partial elements as whole, through closure. Even a single linear sequence of simple images can be perceived as whole shapes despite any obscurity that may be present (see fig. 2.2). While only the arrow and circle shapes are fully visible, we are able to perceive and extrapolate the other shapes in the sequence as a triangle, square, and star. Despite the use of different colors, borders, and shading techniques, we are able to perceive closure of each individual image. However, when a partially obscured object fails to provide sufficient detail, our perception may have difficulty in classifying it. For example, consider the black square-like shape in figure 2.2, which is partially obscured by the circle adjacent to it. Although we may not be able to see the full right edge of the shape, our perception may identify it as one of many things: possibly a square or rectangle, but mostly likely a quadrilateral. Despite any challenges to our perception, the principle of closure suggests an innate desire to form conceptual wholes from information environments,

Figure 2.2. Illustration of the principle of closure. When objects or shapes are partially obscured, if enough details are visibly present, our perception can interpret them as wholes. *Source*: Created by the author.

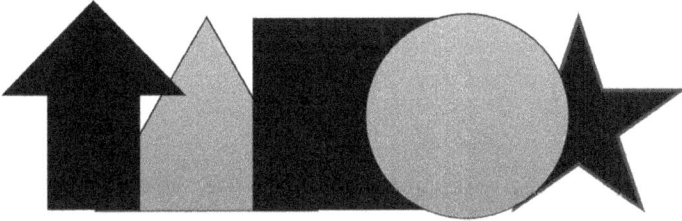

helping us construct meaning. While the actual obscured objects may appear differently, this closure affects how we ultimately perceive and classify each object in our visual field.

The principle of continuation explains how we perceive patterns as continuously repeating beyond the range of our visual perception. Koffka (1935) suggests good continuation supports our perception of stable structures and homogeneity, which are often present in repeating patterns. Seeing a partial pattern may be sufficient for us to perceive it as a repeating element that continues across or down the page, as in a solid border, or into the distance, as in a repeated skyline. When viewing a landscape, we look at the horizon in the distance seeing where the landscape meets the skyline and perceive its continuation despite the limitations of what we actually see. Although our visual senses have distance limitations, our perception helps us conceptualize a fully realized horizon or skyline. In this sense, the principle of continuation relates to both closure and repetition, where we perceive something partially visible and repeated as a complete whole. In an information-based environment, pages might use decorative borders for backgrounds that use repeated elements, which suggest a conceptual continuation. Even the use of negative space can frame or enclose content into individual sections on a viewable screen, which repeats and continues in predictable patterns throughout an entire document, as users scroll down the page. As such, the principle of continuation supports our understanding of how information environments, through their arrangement and organizational patters, form conceptual wholes. Continuation can enhance our information experience by simulating repeating patterns, which help us perceive, read, and scan content more effectively. To illustrate, when we see a series of shapes with similar characteristics and spacing, such as a horizontal arrangement of squares, we are able to perceive this arrangement as a repeating pattern (see fig. 2.3). Despite the apparent fading of squares into the left and right margins, this sequence is perceived as a continuation in both directions. This technique might be useful in creating borders, backgrounds, horizontal rules, or other such features within a page design, without having to create more complex design patterns to achieve the same effect. And whereas the principle of closure assists in the perception of faded squares as wholes, the principle of continuation allows us to conceptualize the sequence as a repeating pattern, working with other perceptual acts, helps us perceive depth beyond the limits of our visual perception.

The principle of figure/ground describes both object and surface, including the space between both, and how collectively they form a complete, conceptual whole (Koffka, 1935). Together, figures and grounds help us understand characteristics such as distinction, function, meaning, and relatedness in information environments. Koffka describes figure and ground as object and framework, which naturally support each other and can be perceived as functioning both separately and together. Figure/ground explains how we perceive an environment through the examination of objects present in both the foreground and background of an environment. Understanding figure/ground relationships helps us determine both relatedness and disparity between objects, such as how multiple groupings may suggest one or more combined, functioning units. This principle underscores the importance of both depth and space in visual perception, where objects can overlap or appear as different layers within a single information environment. Figures are typically perceived objects present in the foreground, while the ground comprises the background

Figure 2.3. Illustration of the principle of continuation. Our perception can perceive repeating patterns as continuing beyond what we can see, such as on the margins of a page. *Source*: Created by the author.

in which objects are placed. When simple shapes are positioned in the same space, we perceive them as having depth, with independent layers, in a figure/ground relationship (see fig. 2.4). Each individual object may be perceived differently, based on their color, distinction, position, shape, size, or other distinguishing characteristics. Within the three images, we may initially perceive shapes as individual objects and then any perceived conceptual groups we may discern. The first image is a right-pointing grey arrow (figure) positioned on a black square (ground). The second image is a black star (figure) positioned on a grey square with a thin black border (ground). The third image is a grey square (figure) positioned in the center of a black circle (ground). These three images can be perceived as separate units or conceptual groups that have both depth and independent layers (a figure and a ground). Typically, elements positioned in the center will be perceived as figures, while larger images with borders that extend beyond each figure will be perceived as ground. However, depending on differences in each object's characteristics, both figure and ground could be discerned quite differently.

Conceivably, the concept of depth within a figure/ground grouping can include multiple occurrences of both figure and ground, where multiple layers are superimposed upon each other in the same space. While our perception can process independent figure and ground elements in a single space, our field of vision is typically quite narrow, only allowing us to focus on the finer details of one figure or one ground at any given time (Arnheim, 1997). However, our perception also processes figure and ground elements iteratively, cycling through multiple perceptual tasks to make sense of the whole. As another example, a painting of a bowl of fruit may be perceived as separate objects, such as apples, oranges,

Figure 2.4. Illustration of the principle of figure/ground. Our perception interprets visual groups of objects or images in the same space as independent and interdependent figure and ground elements, which supports both conceptual and semantic understanding. *Source*: Created by the author.

bananas, kiwis, and the bowl. Each piece of fruit and the bowl may be perceived as figures, which overlap in close proximity. There may also be ground elements, such as a table, painted wall, or other background elements in the painting, which form a complex background. While we examine and perceive each figure and ground element independently, we form conceptual groups to help us better understand the meaning of the whole composition. The principle of figure/ground is also dependent upon other perceptual acts and processes such as closure, contrast, continuation, layering, and spatial perception to help us fully process an information environment. From a spatial perspective, the layering of figure and ground elements affects how we interpret their relatedness, disparateness, and overall holistic meaning. In more complex visual and spatial information environments, each figure and ground may be discerned differently, based on other objects and their perceived groupings.

The principle of proximity refers to our ability to form concepts based on the spatial distances of objects, whether it suggests disparity, relatedness, or similar characteristics, functions, and forms (Kohler, 1947). While the principle of proximity suggests the importance of spatial characteristics, space can also be communicated using visual or textual codes, such as the use of background, color, shading, or textual styling, which support spatial coding through the use of negative space, margins, and gutters to arrange or group content. Proximity is also related to how we perceive object groups, whether elements are layered, overlapping, or separated, and how these groups are attributed with specific meanings or intentions within an information environment (Kohler, 1947). For example, while two visually similar images appear in close proximity on the same page or screen, such as individual shapes positioned side by side, these objects can be perceived as both individual objects and matching groups. Based on the specific visual, spatial, and textual codes used, such as the use of borders, dividers, shading, or spatial distance, we may attribute different meanings or relationships based on their positioning techniques (see fig. 2.5). While a simple vertical bar may separate two shapes, such as a star and a circle, the use of a pipe symbol between them may conceptually suggesting different choices or options. While the three shapes (star, pipe, circle) are placed in close proximity, our familiarity with the pipe symbol representing an either/or choice may change our perception of how space is used to group these elements. Instead, in such an arrangement, the pipe symbol may suggest distance or disparity between the star and the symbol. Similarly, a spatial configuration of five stars positioned equidistantly in

Figure 2.5. Illustration of the principle of proximity. Proxemic techniques such as visual grouping and separation can be perceived as having conceptual similarity or dissimilarity, in terms of features, functions, or relation. *Source*: Created by the author.

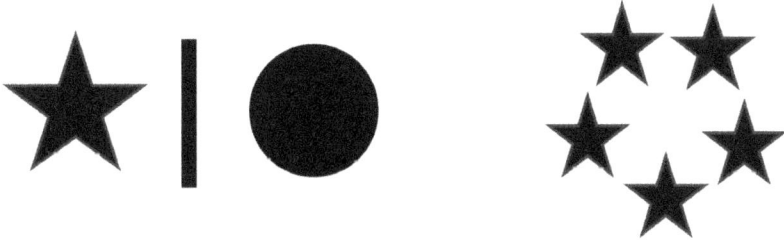

a pentagram configuration might suggest a specific concept, such as a constellation or rank insignia, depending on factors such as the context of use and our own familiarity with similar groupings.

Proximity can convey semantic meaning, whether by close arrangement of objects to suggest likeness or relatedness, or by distance between objects to suggest difference or distinction. A more distant arrangement or separation of objects may suggest greater conceptual differences, such as distinct items in a list or images positioned equidistantly for comparison. Regardless, the principle of proximity describes how space helps us perceive conceptual groups from objects present in a physical space or information environment. In a sense, we perceive semantic connections between objects and the space in which they are placed, supporting the basic acts of concept formation and comprehension within an information environment.

The principle of similarity describes the perceptual process of interpreting objects with like characteristics as consistently similar. Koffka (1935) suggests that when we observe consistent features between objects, we perceive their similarity in both concept and framework. Similarity helps us form conceptual groups or clusters from objects and environment, which supports comprehension. Similarity can be perceived based on both visual and spatial characteristics—whether objects have similar colors, dimensions, positions, shapes, or styles. Similarity helps us classify objects based on their shared characteristics, which can fit one or more categories or groups that have specific meaning to us (Koffka, 1935). Perceived similarity facilitates pattern matching and sorting of objects into conceptual categories, which can be new or previously experienced. For example, a Granny Smith apple might be conceptually sorted into multiple conceptual categories such as apples, green objects, fruits, and others. As a perceptual process of categorization and sorting, the principle of similarity helps us create perceptual concepts, categories, and classifications of objects we observe in simple, as well as highly complex visual-spatial information environments (Baehr, 2010). For example, a gaming computer keyboard may include keys that use similar shapes and familiar icons, such as directional arrow keys, which suggest navigational options for up, down, left, and right, which share similar characteristics and recognizable concepts for experienced computer users (see fig. 2.6). While each arrow key may be visually similar in terms of style, however, the direction or position of each arrow may suggest independent functions. When these arrow keys are grouped together on a keyboard, they are perceived as a conceptual group, with a similar purpose—as movement or navigational controls. If these keys were styled differently or appeared in different locations on the keyboard, their collective meaning and function might be more difficult to discern. The stylistic similarity and grouping of other keys on the keyboard may also suggest like functions, such as grouped

Figure 2.6. Illustration of the principle of similarity. When our perception interprets objects or images with like characteristics as a group, such as different directional arrows clustered together on a computer keyboard, they may be perceived as conceptually (or functionally) similar. *Source*: Created by the author.

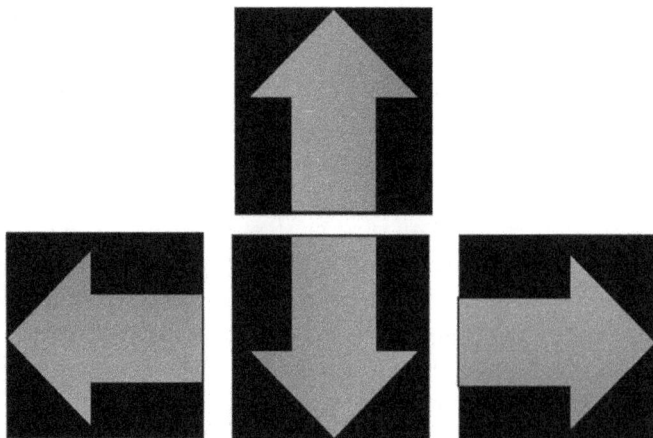

function keys, labeled F1, F2, F3, F4, and so forth. These function keys are typically grouped together on a keyboard; however, their independent functions are often customizable and different based on the information environment in which they are used, such as a game or word processing software program. While similarity may help us perceive conceptual groups or independent functions, we may rely on other perceptual processes to determine specific meanings and uses. Perceived similarity also supports concept formation and overall comprehension of objects and environments, whether they are physical, hybrid, or visual in nature.

Collectively, these principles of Gestalt theory explain how basic human perception functions collectively, as a series of perceptual acts that can adapt and function in both physical and information-based environments. These perceptual processes are also a critical part of how humans form a holistic sense of an information experience. While each individual principle describes different perceptual acts, they are not mutually exclusive, but rather, they function collectively and iteratively, helping us form conceptual wholes, or a comprehensive understanding of objects and environments. Perception governs what we observe and recognize, influences what we see, how we think, and, subsequently, how we make meaning. Therefore, Gestalt principles can be applied to frameworks and guidelines used to design and develop information products and environments, using a combination of visual, spatial, and textual codes. While much of Gestalt theory was written and conceived when information products were largely print-based texts, their application continues to be successfully applied in digital, electronic, interactive, and virtual information design and development practices. Additionally, the principles of Gestalt theory have been applied throughout a range of disciplines and information environments, from the creation of artistic compositions to the development of technical information products. Ultimately, our perception governs more than what we see visually, extending to how we also think, both visually and spatially, within these information environments (Johnson-Sheehan & Baehr, 2001). As such, our perception is an important part of a larger interpretive process of thinking and conceptualizing a holistic information experience.

Visual and Spatial Thinking

What we perceive in an information experience is more than just what we see; in electronic information experiences, it can be augmented by other senses, such as the senses of hearing and touch. Video content can incorporate audio content that enhances and supports communicative messages. Haptic devices allow us to navigate, reposition, and reorganize information through touch-based interface controls. These added sensory inputs can even simulate an experience, through the use of recorded sounds or physical motions that might be present in a real environment. From Gestalt theory, we understand how human perception functions as collective acts, which help us perceive information environments and their visual, spatial, and textual codes. But perception is more than the simple processing of sensory input. When we perceive information environments, both simple and complex, as conceptual wholes, these interpretative processes demonstrate active thinking. The work of Gestalt theory underscores the importance of perceptual thinking in the principles previously discussed. Building upon this work, Rudolf Arnheim (1997) describes these collective perceptual processes as active visual thinking, or the essence of visual perception, which has important implications to our cognitive processing.

To describe how perception functions in this way, Arnheim developed principles of visual thinking that explain how we interpret, think, and act upon our perceptual instincts. Largely, his works focused on examples from art history and theory to explain and illustrate how visual perception operates. While his work on visual thinking was published prior to the onset of the World Wide Web and much of electronic publishing, his principles of visual thinking have applicability well beyond print-based information products (Johnson-Sheehan & Baehr, 2001). Regardless of the medium or environment, perception helps us discern and formulate concepts, solve problems, and comprehend structure, meaning, and intent. These visual thinking principles are extensible to electronic and web-based environments, because these product environments emphasize visual and spatial aspects of information design (Johnson-Sheehan & Baehr, 2001). In particular, the core visual thinking principles, which are applicable to our perception of information environments, include *vision is selective, fixation solves a problem, discernment in depth, shapes are concepts*, and *complete the incomplete* (Arnheim, 1997). Other principles in his collective works include *perception takes time, exploring the remote*, and *how machines read shapes*, which have more cognitive and experiential learning applications. Collectively, these principles describe the processes of how humans acquire, select, and derive meaning from information environments, through related perceptual acts. Accordingly, these principles work symbiotically along with our cognitive processes, which help us construct and conceptualize an information experience.

The first of these core visual thinking principles, *vision is selective*, describes our innate perceptual process of actively selecting and exploring information within our field of vision (Arnheim, 1997). Our visual perception is deliberate, focusing on specific elements, often one at a time, sometimes over and over, depending on the complexity, familiarity, and importance of what we see. For example, we prioritize our focus on objects with higher contrast, distinctiveness, motion, or uniqueness, and then focus on those less so, following a visual hierarchy of sorts, from most to least distinct or novel (Baehr, 2002). In a website, we might notice animated banners, short video clips, interactive navigation tools, and images of greater contrast before we notice other visual features, such as the heading levels or font faces used for standard text. What we notice first may also depend on the unique position or location of an element on the page, suggesting the importance of spatial characteristics in our visual thinking. This active perceptual selection and exploration determines how we prioritize information for further study and analysis (Arnheim, 1997). While our range of focus may be small when studying a single object at close range, our perceptual focus shifts as we explore other characteristics, features, and objects present in an information environment. This process of perceptual prioritization and selection also forms the basic response upon which other visual thinking principles are applied.

When viewing a visual image, our perceptual processes are innate and engage almost immediately. While a single visual element, such as a painting or depicted object on its surface, may capture our attention and serve as our initial primary focus, we may undoubtedly notice other independent elements in the same visual space, including the framing, lighting, medium, and other characteristics, which nevertheless contribute to our perception and overall information experience. As an extended example of how visual focus and perception function, consider a single image, such as a watercolor painting (see fig. 2.7). This image depicts a large open space, such as a crowded antechamber, foyer, or hallway, adorned with various forms of lighting, signage, windows, and people. Initially, our focus may be drawn to notice higher contrast elements, such as hanging lights, illuminated signage, or darkly framed windows. Our visual perception may focus on specific elements portrayed in the image, such as a facial silhouette, a small family, or even, eventually, the figure in the foreground, which appears to be staring at us from afar. Noticing one person in the crowd may lead our focus to shift toward other persons and their individual differences. Depending on our level of engagement and interest, we may spend a significant amount of time studying messages, signs, or symbols depicted on the hanging banners in the upper left-hand corner. We may also notice lighting differences on the right-hand side of the image, whether they appear to be hanging illuminated strands or longer, rectangular sconces positioned between each large window. While our individual interests and motivations may change and differ, our initial perceptual response innately focuses on its visually distinct elements, starting with those with the most brightness, contrast, emphasis, or even relevance to our interests. Gradually, our perceptual focus will shift toward other elements, perhaps those with less distinction, as we study other elements in the composition that are worthy of further attention. While the perceptual principal *vision is selective* governs this process of selecting visual elements for study, it operates symbiotically with other perceptual acts as part of our holistic process of visually thinking. The selection of visual elements and focus is, in essence, the collection and initial perceptual processing of visual stimuli, which eventually leads to more complex perceptual and cognitive tasks. As we begin to perceive other characteristics of the image, such as the semantic relationships between objects, the specific meaning of shapes and symbols, and so forth, our perceptual acts help us process a single image as a more complex and interactive visual environment.

The principle of *fixation solves a problem* describes the perceptual process of identifying elements in our visual field that address a specific information need, such as completing a task, process, goal, or desire (Arnheim, 1997). The act of fixation relates to the previous principle of *vision is selective*, in that what we choose to examine is often deliberate and may be selected for further examination, or fixation, because we believe it may be useful in solving a problem or information need. Fixation is somewhat different from focus, in that it suggests a longer duration of perceptual focus, beyond an instantaneous gaze. While we may examine multiple objects in an information environment, we may fixate specifically on ones that are relevant to a current or immediate information problem or need, which may help us complete tasks or comprehend meaning. Fixation may also help us understand a concept, function, organizational pattern, and relational or semantic characteristics. As a result, active fixation is often based on perceived relevance or usefulness. For example, we may fixate on a navigation menu or keyword search in a website to help find specific

Figure 2.7. Image of a night at the Kennedy Center. When viewing a complex image or visual composition, our perception actively selects elements for examination and follows a hierarchy from those elements most distinct or relevant to those less so. *Source*: Liz Pohland, *Night at Kennedy Center* [Watercolor], 2020. Used with permission.

information (Johnson-Sheehan & Baehr, 2001). These navigational tools combine visual, spatial, and textual elements, such as the use of text labels, boxes, buttons, hyperlinks, and other elements, which aid browsing and searching tasks. Once we learn and perceive a particular tool as useful, we may rely on it for more complex or subsequent problem-solving tasks. Ultimately, as we fixate to solve problems, we prioritize what we perceive to be important for a particular purpose or need and eventually move on to the next object we find relevant or useful.

Fixation implies that we have identified elements or objects within our visual field, which require special attention to satisfy a specific goal, such studying its unique details, features, functions, or purpose. We might fixate on a specific object because of its abstract characteristics, distinctiveness, novelty, or even perceived usefulness. Fixation would seem to also imply motivation on some level, whether specified or unspecified, such as a simple reaction to something new or out of place within an information environment. For example, when accessing the home page of a website, we might fixate on a nested navigation toolbar to discern how it functions, or even how it might help us better understand the overall organization of the site as a whole. However, we might also focus on an animated flashing symbol, which seems to appear and disappear randomly in the upper right-hand corner of the page, simply because of its novelty. While our perception would seem to be rather directed in one instance, it can also function almost instinctively in others. Instinctive fixation might even lead to a more purposeful and directed activity, once its perceived value is realized through such acts.

The principle of *discernment in depth* suggests our perceptual focus continually shifts between objects in the foreground and background details as we explore in an information environment (Arnheim, 1997). While our perception focuses on objects independently, the acts of exploring foreground and background are mutually exclusive perceptual processes, not simultaneous ones, which are integrative and interdependent as we form a conceptual whole (Johnson-Sheehan & Baehr, 2001). Discernment in depth suggests the iterative nature of perceptual focus, which through repeated examination of figure and ground, refines our understanding of the whole. For example, in landscape photography, we may study objects of interest in the foreground, such as animals, people, structures, or other figures present, and switch to studying the background, such as the colors in the sky, the position and shape of clouds, the setting sun, or other details. Consequently, our perception is adaptive, allowing us to quickly move back and forth between objects of interest in the same environment, which help us discern specific characteristics in each layer of depth present in the image. While visual codes such as color, contrast, shape, size, and style are some of the important details perceived by our senses, the principle of discernment in depth also underscores the importance of space and the relationship between objects in the same environment. Some of these spatial codes include depth, dimension, distance, order, and position, which together with visual codes help us form a holistic sense of the entire landscape. As a result, the combined codes present in an image (or information environment) help us discern levels of depth, which are perceived individually and as a collective whole.

Discernment in depth is closely related to the Gestalt principle of figure/ground, which suggests the importance of both objects and environments (in which they are placed) in design. Depth also suggests multi-dimensionality within information environments, which are composed of multiple layers of figure and ground elements. A photograph of an object, such as a vase of colorful flowers displayed against a white background, nonetheless is perceived as having multiple layers, despite any techniques used in its composition that might suggest otherwise. And despite the apparent lack of styling in using a plain white background, which may suggest a minimalist technique using negative space, this independent layer, or ground, is perceived in terms of its limited characteristics, as well as its relation to the photograph as a whole. Discernment in depth can also be extended beyond image perception to information structures, such as those found in websites (Johnson-Sheehan & Baehr, 2001). In much the same way, information levels of a website may include a home page, several second-level node pages that represent major content sections, and even subsequent level pages of more detailed content that support each major section. When accessing any single page within a website's information structure, our perception actively examines its characteristics, but also its relationship to the whole site. As we visit other pages within the same site, each page, or content layer helps us understand the semantic relationships present between pages within the site. This process of discerning the information structure, its individual layers, and overall organization of content pages within the site demonstrates perceptual visual thinking. In much the same way we discern depth within an image, our perception also discerns depth within information structures, such as websites and other information environments.

The principle of *shapes are concepts* focuses on how we perceive objects and their specific meanings within various contexts (Arnheim, 1997). These objects can be boxes, borders, icons, images, patterns, shapes, symbols, and so on, both abstract and concrete. *Shapes are concepts*, as a visual thinking principle, suggests that we actively sort what we see into classifications of discrete visual categories (Arnheim, 1997). We may recognize some objects as concrete or familiar representations, but we may also classify them as abstract or unknown if we are unfamiliar with their range of features, use, or meaning within a specific context or environment.

Once processed, these shapes and their categories form the basis of a visual perceptual library of sorts, which can help us process the meaning of other objects we encounter in information environments. For example, when learning the basic functions of a software program, we may associate each button, icon, shape, or symbol with a specific meaning, based on familiarity or other codes present, such as textual descriptors. Within a wiki-based environment, a question mark symbol might represent a help library, an hourglass might suggest a search function, or a left arrow icon might link back to the previous screen or even a starting point. As we associate specific concepts or functions with objects in this environment, we begin to perceive details about their independent function, meaning, and relationship with other objects present. While many software programs may incorporate the use of shapes or symbols that seem universal in nature, they can be interpreted differently depending on a variety of factors, including our individual (and previous) experiences, unique product contexts, or other organizational conventions that govern their function, meaning, or use. Therefore, this principle suggests that our perception of shapes includes semantic coding, which may be heavily context dependent, based on any number of factors. While our prior experience may assist us with interpreting some shapes easily, our initial perceptions may be challenged by their use in new and different ways.

However, shapes may not always be concrete figures such as objects or symbols, but may represent backgrounds, grouping strategies, or even patterns within an information environment. While a shaded background may represent a familiar shape, such as a rectangle, our initial perception of it as a four-sided geometric shape may be less important than its actual function, such as a strategy for grouping units of content within a confined space to suggest relatedness. While this technique may be a common page layout technique used in a wide range of information products, such as instructional manuals, technical descriptions, and websites, the perceptual challenge often lies in determining the semantic relationships within grouped regions. Similarly, a series of small, semiopaque circles equidistantly spaced around the margins of a page may be perceived as a decorative yet subtle page border, rather than a series of repeated shapes with independent meanings. While the relationship between these shapes is somewhat simple, the perceptual complexity lies in determining their collective, conceptual meaning. While our perceptual library of shapes and symbols may provide a set of baseline characteristics for objects we have encountered, permutations in use may challenge this perceptual intelligence, of sorts, as we attempt to form new concepts from the configuration of shapes and symbols we encounter.

The principle of *complete the incomplete* describes the culmination of our perceptual processes, to construct a holistic understanding of an information environment. Kohler (1947) argues that, despite our ability to analyze individual objects and their characteristics, the ultimate goal of our perceptual processes is to form conceptual wholes. This principle also relates to the Gestalt principles of closure and continuation, in that we strive toward forming complete pictures, often from incomplete elements in an information environment. For example, while looking a single infographic presentation comprised of multiple bar charts, pie graphs, and tables, we perceive conceptual wholes based on different categories of data presented, as well as the entire presentation itself. Ultimately, these smaller conceptual wholes or categories help us form larger ones as part of a more in-depth constructed understanding. This process cycles through a series of accepted and rejected concepts, through iterative and overlapping perceptual and cognitive processes that help refine our understanding of the whole (Arnheim, 1997). As another example, a jigsaw puzzle, comprised of many unconnected individual pieces, slowly becomes individual sections or smaller wholes, such as frames or recognizable images as we connect various pieces, until, finally, we form a conceptual whole from completing the puzzle.

Similarly, our perceptual processes work toward completing a larger conceptual puzzle from the individual conceptual pieces present in an image or environment, such as in the painting of a school of fish shown in figure 2.8. The spacing and lighter shading techniques used in the image would seem to suggest a puzzle-like configuration, where different abstract and concrete shapes would seem to fit closely together to create a singular, holistic image. Our perception of differences in color, contrast, shape, and spacing help us perceive individual fish within the image, perhaps based on differences in color or the positioning of different eyes throughout the image. While these perceived differences may help us identify different fish within the image, we also are able to form conceptual groups, such as lighter or darker fish, smaller and larger fish, or even an entire school of fish swimming together in close proximity. While the school of fish occupies the entire image, our perception may even extrapolate other environmental characteristics and attribute them to the image. We may perceive (or even expect to see) an aquarium or an ocean environment in the background, due to our perceptual expectations from similar spaces. Using familiar shapes, such as individual fish, and close proximity and spacing between them may contribute to this perception, or even others we've experienced previously. Collectively, our perception helps us determine concepts, contrast, difference, position, similarity, and other characteristics within the image, which formulates a conceptual whole, conceived from our perceptual acts, and contributing to our overall information experience.

Collectively, the principles of visual thinking demonstrate how our perception functions within a vast range of information environments, whether they are applications, artwork, books, documents, information graphics, websites, or other information environments. Our visual thinking governs our perception of physical objects and electronic and virtual environments in similar ways, as we examine their unique visual, spatial, and textual codes. In particular, electronic information environments have unique features, including their ability to deliver interactive navigation

Figure 2.8. Image of a school of fish swimming in water. Collectively, our perceptual processes form a conceptual whole from the disparate elements and features present in a visual composition or environment. *Source*: Mike Fuller, *MultiGolds* [Acrylic print], Society6, https://society6.com. Used with permission.

menus, keyword search, customized content, enhanced visual coding and styling, hyperlinking, interactive media, layered or multidimensional content, interactive applications, and responsive design features, among many others (Baehr & Lang, 2019). Regardless of the environment and its characteristics and features, our visual thinking readily adapts to these environments, enabling us to construct meaningful information experiences from them. Information design practices and design principles widely used in technical communication and other fields are informed by theories and principles of human perception, which include Gestalt theory and visual thinking. When we understand how users focus, solve problems, discern depth, interpret shapes, and form concepts in these environments, this information can be useful in designing effective products, environments, and information experiences.

When Perceptual Processes and Information Experience Misalign

Sometimes, information products are designed and developed in ways that are a mismatch for our perceptual processes, creating experiences that are unexpected or even frustrating for users. While our perceptual processes may help us focus, fixate, discern, conceptualize, and comprehend various elements present in an information environment, sometimes our perception can become overwhelmed when confronted with confusing or unfamiliar stimuli. For example, when presented an excess of information simultaneously, whether in the same communicative mode, media type, or in different ones, this information overload can create disinterest, frustration, and negative impressions about an information product. When our perception fails to interpret simple concepts, functions, or meanings of elements, it can also impact other aspects, such as product credibility, individual performance, and task completion. As such, it is important for designers to fully understand how perception affects perceived experience and how to apply effective design tactics and techniques, which support user perception and information experience.

Arnheim (1997) suggests that our vision is selective, as one of his visual thinking principles, which explains how we focus and prioritize visual objects based on their distinctive characteristics, emphasis, uniqueness, and usefulness relative to other objects in the same environment. When an information environment incorporates too much visual clutter, such as excessive use of animation, contrast, repetition, shading, or other elements, this can confuse our focus and overwhelm our visual senses. For example, some websites may incorporate poor use of color contrast, inadequate spacing between visual elements, excessive use of small serif font faces, cluttered background images, or other styling problems, which can create too many visual elements competing for our attention, making it difficult to focus on items most relevant or useful. Design techniques that fail to account for perception in their application can lead to poor comprehension, use, or even usability. Conversely, websites that fail to emphasize or highlight important elements or features can create similar focus problems for our perception. Therefore, balancing these techniques can be important to help users interpret visual elements as a hierarchy, from those important and purposeful to those that are more subtle and thematic, by employing proper use of design techniques that incorporate visual perception.

The perceptual process of selective focus naturally leads to another visual thinking principle, fixation solves a problem, which helps us interpret visual elements in terms of their function and usefulness (Arnheim, 1997). Focus is about perceptual recognition, while fixation is more about performance. When we fixate, our perception spends additional time and effort interpreting a specific element to determine how it might be useful to our current action or motivation. Sometimes these functions are easily determined, such as using a keyword search or hyperlink to access information, or by sorting data in a spreadsheet by columns or rows. However, when these simple functions are improperly labeled, placed, styled, or tagged, they might be misinterpreted as having different functions or uses than intended. Abstract or unfamiliar objects, such as icons or shapes, may be improperly perceived in terms of their function, particularly when there is insufficient instruction or contextual information present. For example, navigation menus or tools in websites may use icons that are a mismatch for their actual function or meaning. While the symbol of an hourglass may suggest a search function in a website, a microphone symbol may not always carry the same meaning for users in a virtual meeting space as it does in a video recording application. Since our understanding of these functions and tools often depends on context, it is important to include appropriate visual, spatial, and textual codes that support the intended function or meaning of objects, which aid both understanding and performance for users.

Understanding the characteristic differences and functions between figure and ground in information environments is an essential perceptual process. When our perception encounters new environments, our combined perceptual processes focus and fixate on both figures in the foreground and ground elements present in the background, but not simultaneously, as we discern depth within an information environment (Arnheim, 1997). Discerning figure and ground elements helps us interpret the meaning of conceptual pairs, functions, patterns, and relationships between various information layers that are present. Sometimes, these elements are miscommunicated or misinterpreted based on problems with their positioning or styling, or even because of their presence in an information environment. Figure and ground elements are often perceived as groups of conceptual pairs, where one often supports the other. For example, a product information description may include both foreground elements (images, logos, slogans, taglines) and background elements (colors, decorative borders, thematic images, watermarks) that function together to create a unified brand or message. This unity of figure and ground may also be supported by the textual descriptions in the product description. In some cases, a standard template may be used for the product, which may be less visually appealing or organized, which lack complementary figure and ground elements for ease of comprehension. Some figure and ground elements may also be meant to communicate disparity, rather than similarity, which are just as important to how we discern the relationships between information layers. When selecting and placing figure and ground elements, it is important to consider their potential perceived relationships and meanings to avoid potential confusion or discernment problems, particularly in more complex information environments.

Our perception also helps us interpret objects, symbols, and their meanings within an information environment, which is suggested by the principle of *shapes are concepts*. Shapes and symbols can help us understand abstract, complex, or even unfamiliar concepts, depending on their use within a particular information environment and context. Shapes can be symbolic, such as icons or images, or even be used positionally, as borders, background, or shaded areas. When paired with textual messages, shapes can suggest a complementary meaning to help with comprehension and use (Johnson-Sheehan & Baehr, 2001). For example, website navigation toolbars that pair icons with textual descriptors help users more clearly understand the function of each toolbar item. Pairing a question mark with the word *help* might suggest assistive tools such as troubleshooting tips or frequently answered questions. Similarly, pairing a house icon with the word *home* might suggest a clickable hyperlink that accesses a home page. Sometimes when unfamiliar symbols are used without contextual cues, such as text descriptors, users may misinterpret their intended function or meaning. As an example, a simple luggage icon used in an airline website may have multiple meanings, such as baggage, business travel, carry-on rules, flight schedules, and so on, while no singular meaning may be obvious to new users. While some shapes may have familiar meanings based on previous experiences, unfamiliar ones may create potential confusion without the proper contextual details. As a result, it is important to consider how shapes create meaningful concepts, which users recognize and rely on to perform basic functions with information environments.

Working together, our collective perceptual processes work in tandem to help us form a complete understanding of an information environment, which is described by the principle of *complete the incomplete* (Arnheim, 1997). While our perception forms a complete understanding from the various codes and objects present in an information environment, any miscommunicated or missing elements can lead to comprehension and performance problems in a user's information experience. For example, when we fail to understand the advanced search tool in an online tutorial, this experience may cause us to associate other negative aspects with the information product itself. As a result of these problems, we may interpret the collective information experience to be something quite different, or even diminished from what might have been intended. Whether focusing, fixating, discerning, or conceptualizing, our perceptual processes affect our interpretive experience, good or bad. As a result, it is essential to consider how perception may affect how we design and develop information products and environments for optimal comprehension and use.

Implications of Perception and Information Experience

Human perception is inextricably linked to how we process and think in information environments, and our understanding of these processes can help use create more successful information experiences and products. The principles of Gestalt theory and visual thinking explain how our perception prioritizes and processes the various visual, spatial, and textual codes and layers in those environments. The close connection between the visual and spatial characteristics is even more pronounced in electronic information experiences, illustrated through the use of sophisticated information visualization tools, navigation options, and interactive content used widely in information products today. Advanced content authoring tools and methods allow for the creation of more dynamic, interactive, and layered information environments, which demand more from our perceptive abilities. As such, information experiences can incorporate content and design elements that are both static and adaptive, live (streaming) and historical (archived), governed by a vast array of external conditions (positional, situational, temporal) and our own unique individual preferences. Additionally, hybrid and electronic information products often incorporate multiple content layers and modes, such as interactive media, pop-up windows, multiple tabs, and externally referenced (or hyperlinked) content, embedded on the same page or screen. These overlapping layers are often perceived as three-dimensional environments, despite the limited dimensional qualities of a flat page or screen. As an example, from a coding standpoint, developers can create layered content in three dimensions using the Cascading Style Sheets (CSS) z-index property to create layered or stacked elements that appear on web pages. Even content archives, such as the Internet Archive, may include archived versions of content that suggest interactive temporal components to browsing and searching document collections. In turn, these characteristics affect how we interpret content, such as based on its time and date stamp, adding yet another dimension to our information experience. Our perception helps us discern these added layers of information and complexity, enabling us to develop more adaptive approaches in navigating information environments.

Our perception of information experiences depends heavily on content as well as medium. Information products may use a variety of visual, spatial, and textual codes and techniques, which include the use of color, images, perspective, symbology, texture, and the linguistic messages present. These overlapping codes and techniques often support and augment one another, such as when the spatial positioning enhances our understanding of images and their captions. For example, a side-by-side presentation of images and titles of products on a web page may suggest a comparison for users to discern both similarities and differences. Consequently, the integration of spatial codes, or positioning techniques, can alter our perception of the visual and textual messages present. In the previous example of a painting, the background elements present suggest spatial characteristics through the painting's frame, depth, lighting, matting, positioning, shape, and so forth, as well as any characteristics present in the composition itself. Our perception also helps us understand the relationships between the various unique elements, which together contributes to our collective information experience, whether we're interacting with a document, image, physical object, or virtual environment.

While our interpretation of an information environment is shaped by our perceptual processes, other important factors contribute to information experiences. Our collective information experience is shaped by both the user and the developers of information products and environments. For example, developers may plan products with deliberate characteristics and features, such as interactive navigation menus, complex grid layouts, colorful style sheets, and personalized content, which are intended to communicated specific messages to users. Whether successfully communicated through these intended features, our perception ultimately determines what conceptual whole we perceive as an information experience. Ideally, these two perspectives—the user's perception (or conceptual whole) and the developer's intent will align in more ways than not. Accordingly, there are important implications to consider to ensure that design practices align with how users perceive, think, and interpret information products and experiences, holistically.

Human perception forms the basis of our visual thinking. Our perception is an integral process, governing how we see and think, which is inextricably linked to our cognition and learning (Arnheim, 1997). Gestalt and visual thinking principles explain how our perceptual processes function holistically and suggest that users are thinking visually and spatially in both physical and information-based environments. Whether viewing a painting or an interactive web application, users perceive characteristics such as space, dimension, and perspective in some ways similarly, but in other ways quite differently. Virtual environments may simulate physical characteristics and dimensions, which enable us to understand them by making connections to objects or environments in the real world. Whether objects appear to be placed on top of each other, close by, far apart, or layered in the same environment, these spatial characteristics are also perceived in terms of their function, meaning, and relation to other objects as a complete, functioning whole. As a result, there are fundamental differences in how we navigate, interact, use, and ultimately comprehend an information environment. As another example, when shopping for groceries on a website, we might first notice product images, brand logos, animated banners, or pop-up messages on specials that appear on our screen. We may also notice how these elements are positioned spatially to help us in determining product categories, similar products, or the like. On an individual product page, we might notice a grouping of images that show different views or groups of related products. Navigation tools that help us browse and search products that we perceive may perceived in terms of their function or location. What this suggests is that the act of perceiving these elements demonstrates active visual thinking that helps us determine function, meaning, use, and much more.

The principles of visual thinking are overlapping processes that holistically govern our perception. Our visual thinking involves an iterative series of perceptual acts such as focusing, fixating, discerning, and conceptualizing, which can be instinctive, interdependent, and even symbiotic, helping us learn more about individual objects and their holistic environments. Our perception may initially involve focus, but it may quickly, or instinctively, move to fixation or concept formation to study specific objects, but in greater detail. Depending on interest and motivation, our perception may engage other acts, such as dissecting and discerning the different layers within a visual environment. Visual thinking encompasses the full range of these discrete acts; however, our perceptual responses may change over time, based on our own experiences with different information environments. For example, an animated high-contrast flashing advertisement on a website, may initially gain our visual focus or even warrant additional perceptual study; however, subsequent exposure to these ads may desensitize our interest and over time, we may simply ignore similar presentations.

Perception is adaptive to a wide range of information contexts and environments. As our perception actively works toward the formation of a conceptual whole, its processes are adaptive and iterative through multiple cycles of concept formation, rejection, and refinement. Initially, this process can result in a stable (somewhat temporary) conceptual whole but depending on subsequent changes or interactions with the same object, may eventually, and perpetually change. We may focus on the same object in a single setting, studying it feature by feature over multiple iterations before we arrive at a satisfactory and holistic understanding (Arnheim, 1997). This adaptation suggests flexibility in our visual seeing and thinking. We may respond differently when encountering new informational contexts and changes and as a result, our perceptual responses may change over time to better suit our information needs based on different products and contexts. For example, consider the evolution of the cellular telephone into the smart phone. Initially, the cellular telephone functioned primarily as an audio communication device, but with subsequent upgrades in its technological and communication capabilities, its use and our perception of the information experience using it changed. As future iterations evolved, the device became the smart phone, with differences in features and functions, including text messaging, video communication, online researching, application, e-commerce, online publishing, and others. And as a result of this evolution, our perceptual understanding and information experience changed. Throughout changes in information products and environments, our perception remains adaptive, helping us continually refine and update our understanding of a conceptual whole. While developers can rely on perception to adapt to changes in information products, they should also consider including ways that support users in adapting to new contexts and environments.

Perceptual processes can evolve with continual exposure to new stimuli. Our perceptual processes can also adapt and evolve as we learn more from an information environment (Baehr, 2002). An information experience is shared between developer and user, where its characteristics may be integrated by developers into information products, but ultimately that experience is also interpreted by its users. While there may be some disparity between the application, intentions, and understanding of that experience between developer and user, the conveyed information experience will undoubtedly change over time through future product iterations. Consequently, our perception interprets an information experience differently with each iteration and experience with that product and its environment. Our perceptual processes can also change over time, altering our initial reactions to certain information configurations and changes we encounter. For example, our initial perception of a software tool's keyword search function may change after successive uses, as we discover new shortcuts and workarounds that better suits our needs. We may find the tool's autocomplete feature useful in locating related topics that appear while typing, without executing an actual search. Once we learn new shortcuts or uses for one type of tool or information product, our perception of other tools may evolve as well. And while the tool itself may remain unchanged over time, our appropriation of it may change based on actual use. As a result, the next time we encounter similar features in other information products, we may perceive and think about their applied uses in different ways than we did previously. As an evolutionary process, perception of that information experience will likely differ in later interactions as well.

Information development practices should consider that this perceptual evolution may impact user interaction and performance when creating information products that best suit the needs of users.

Perceptual principles can serve as useful guidelines and heuristics for information designers and developers. The principles of perception and visual thinking can also be applied and used as guidelines and heuristics for developers. Whether, in a practical sense, we apply these principles to create documents and products or use them as assessment heuristics, such as in user experience testing, these principles can help developers better align information products with their intended users, and subsequently, their natural perceptual processes. In many contexts, perceptual theory forms the basis of specific design principles and best practices commonly used in many disciplines today, including technical communication and user experience. Perceptual principles also help developers understand how users perceive and think in information environments and then develop practices to design and develop information products in more user-centered ways. From a tactical perspective, these principles of design and specific applied techniques are discussed in detail in a later chapter, covering commonly used practices in both information design and user experience design. While these principles help developers create and design useful information products, they must also evaluate and test how these principles are within actual product contexts and constraints to ensure their effectiveness. Using these theories and principles to develop heuristics for testing information products with users can help with this task. As such, when using any set of information design principles, developers must ensure they are properly used and adapted for specific users, purposes, and contexts.

Perception informs both users and developers in understanding the semantics of information environments. When we notice specific characteristics of an information environment, we interpret their meaning as well as the semantic relationships between elements in the same environment, whether a book, software program, interactive game, or virtual simulation. For example, in a coding environment, semantic content can be deliberately communicated through the use of specific element tags that denote an element's relational meaning in the content markup. Hypertext Markup Language (HTML) includes a set of element tags that have semantic meaning as part of their content properties. These include element tags for figures, footers, headers, navigation, section, and titles, among others. Semantic markup can also be used to communicate specific content functions and properties such as descriptions, headings, keywords, titles, and even navigation or searching tools. Furthermore, content can be visually, spatially, and semantically enhanced by the use of style sheet language declarations, such as Cascading Style Sheets (CSS), which apply additional stylistic and positional properties to content, such as depth, emphasis, position, as well as semantics. In much the same way, our perception actively seeks out these semantic codes and relationships, whether they are stated explicitly or implicitly, to better understand information experiences. As such, the visual, spatial, and textual codes used in creating and designing information products helps both user and developer communicate and interpret semantics within an information product environment.

Collectively, the implications of human perception suggest its important function helping users understand and developers construct information products and environments, both simple and complex. Human perception is more than a series of instinctive acts—but is rather a series of complex, iterative processes that demonstrate active and deliberate visual thinking. However, perception is only the first step in a larger process. Our perception also works closely with our cognitive learning processes as a back-and-forth interplay that helps us analyze and comprehend memorable information experiences. While perception is largely reactive in nature, cognition is largely analytical—helping us to refine our concepts of information we encounter. Put simply, what we see and think determines how we act, as the interplay between these processes influences our overall information experience. Working together, both our perception and cognition help us fully comprehend the function, meaning, and uses of information products, including the relational, semantic, and structural aspects of content present in the various visual, spatial, and textual codes of a communicated message.

Chapter 3

Cognitive Experience and Learning

While perception is instrumental to information experience, our cognitive processes enable us to encode, process, and learn from perceived stimuli present in an information environment. In fact, our perceptual and cognitive processes are inextricably linked, which forms the basis of our visual thinking and learning (Kohler, 1947; Arnheim, 1997). This interchange of seeing and thinking leads to our making meaning, as an iterative process of exploration and interpretation of different contexts, modalities, and perspectives of information environments we encounter. Cognition can be defined as the collective acts and processes by which we acquire, appropriate, and use new knowledge (Reed, 2022). Cognition includes our receiving, storing, and processing of sensory information, whether visual, auditory, tactile, olfactory, or gustatory (Arnheim, 1997). Our cognitive understanding of characteristics, features, and semantics of new environments may also be influenced by other factors, such as our current emotions, situational awareness, motivation, time of day, and other relevant states that affect our information experiences.

Both perception and cognition also play critical roles in our visual thinking, helping us to sense, interpret, and act upon new information environments we encounter (Arnheim, 1997). This is particularly evident in electronic information environments, such as online meeting spaces, websites, applications, and other hybrid, simulated, and virtual environments. Our cognitive processes help us decipher, encode, react, and understand various stimuli in these environments. They include the acts of categorizing, comparison, contrast, filtering, pattern recognition, organizing, and sorting, which help us comprehend and learn from information environments. While our perceptual processes interpret the visual, spatial, and textual codes present, our cognitive interpretations (or distortions) may change our actual comprehension of these codes, individually and holistically (Barry, 1997). For example, we may have specific, visceral reactions to seeing flashing yellow traffic lights when we stop at a street intersection, but depending on whether we are driving a car, riding a bicycle, or simply walking in a crosswalk, we may have different cognitive understandings (and subsequent responses) based on the current context as well as our previous experiences. In turn, our cognition helps determine our possible reactions, whether we apply the brakes, stop walking, pause briefly, or continue on our way. Our cognitive understanding may also differ depending on whether the experience is real or virtual. Regardless of the conditions, our cognitive processes can help us adapt to different environments as we learn more from our interactions with them.

Our basic human stimulus-response describes how we respond to new information, but not necessarily our interpretations or processing of that information (Reed, 2022). As we examine the codes, features, and messages present in an environment, we learn more about its specific characteristics, elements, and structure, which contributes to our holistic understanding. Our cognition helps bridge the gap between sensory input and behavioral reaction (Gordon, 1989). This includes what we select to focus on, how long we fixate, what we dismiss, concepts we form, and what we choose to explore further (Arnheim, 1997). Our cognitive abilities help us perform complex tasks, whereby we seek out patterns, similarity, difference, semantics, and priorities in how information is categorized, filtered, sorted, disregarded, and ultimately stored in our memory as learned experiences. Yet our cognitive experience involves both rational and emotional processing of new stimuli, suggesting the importance of our own preferences and situational awareness when interacting with new information environments (Barry, 1997).

Our learned experiences are the result of our combined perceptual and cognitive processing, whereby we attach meaning and significance to new stimuli (Gordon, 1989). Our learned experiences are stored in memory, which can be used to help us adapt, comprehend, and learn from new information environments. For example, when using a new online meeting platform, we may learn its basic functions used to communicate, including audio and text chat, screen sharing, video presentation, and so forth. Our ability to understand the buttons and tools that perform these functions is reliant on both our perceptual and cognitive processes to their various uses. Through subsequent experimentation and use of these tools, we may learn to customize the meeting space to fit our individual preferences and uses of audio, chat, screen sharing, video, and other features to help us communicate more effectively. Even bothersome features may be disabled or ignored, as we learn more from our information experience with the meeting space. As a result, our cognitive learning helps us adapt, appropriation, customize, and communicate more effectively in these kinds of information environments. Through this process of learning, we can even optimize the information experience for ourselves and other participants. Cognition supports learning at different cognitive levels, whether we seek to master new knowledge, form basic comprehension, apply learned experiences, analyze information, or create and evaluate new knowledge (Bloom et al., 1956). Our basic learning aptitudes and modalities, including auditory (speech, music), linguistic (reading, writing), tactile (kinesthetic, touch), and visual (images, space, symbols) also supports cognition and learning (Fleming & Mills, 1992; VARK Learn Limited, n.d.). Collectively, these learned experiences help us process new concepts, differences, patterns, and similarities in a wide range of information environments.

This chapter explores how cognition and learning support our constructive understanding of holistic and meaningful information experiences. It also discusses some of the important connections between our cognitive and learning processes, including relevant theories and practices. Specific cognitive processes and concepts covered include comparison, categorization, filtering, pattern matching, and sorting, as well as different levels and modalities of cognitive processing from basic knowledge acquisition to knowledge synthesis. And finally, the chapter discusses how cognition supports our visual thinking in both simple and complex information environments, and how our learning processes and modalities affect our interpretation and mastery of new information.

How Cognitive Processes Support Comprehension

Cognition involves multiple tasks, processes, and filters (unique to our own experience), which depend on a wide range of factors such as the particular task, situation, and motivation. Cognition studies focus on pattern recognition, categorization, concept formation, language use, problem solving, and decision making (Reed, 2022). While some of our cognitive processes are innate, many adapt and evolve depending on our information experiences. Cognitive processes are not mutually exclusive, but rather they work together and iteratively to help us perceive, memorize, problem solve, and ultimately learn (or encode) new information (Reed, 2022). Cognitive processes can also vary in their complexity, from basic comprehension to higher-level analytical and synthesis processes. Cognition also involves understanding the relational aspects, or semantics, within an information environment. Semantic processing involves interpreting the distinguishing relational characteristics, position or placement, and the unique visual, textual, and spatial codes present that support meaning. This includes how semantic codes communicate similarity and difference, proximity (including closeness and distance), and specific concepts. Much like our supporting perceptual processes, cognition functions iteratively and symbiotically to help us comprehend and learn from new information experiences.

Our ability to recognize and recall patterns is a form of cognitive grouping, whereby we associate items based on similar characteristics and parts within an information environment (Reed, 2022). Pattern recognition helps us match what we observe with other learned and memorized patterns. But even these stored patterns can be revised or replaced with new ones, particularly if no comparable matches are found. Stored patterns help us quickly identify and understand parts, or even wholes, of new patterns we encounter. For example, we may notice different colors, forms, and shapes within a set of symbols, comparing them to other familiar or similar configurations previously encountered. Our cognition may classify or organize these shapes conceptually, by color, form, or even similar shape, as we attempt to match patterns within the set to understand their relatedness. When we see a collection of shapes or objects, we may identify different kinds of patterns within the same information environment (see fig. 3.1). Shapes with similar color (black shapes vs. grey shapes) may suggest one pattern. Other patterns may be observed based on similar shape (arrow and triangle), exact shape (two circles), or shapes with no apparent conformity or similarity (black square and grey star). Some items may belong to multiple groups, or simply be interpreted as unique shapes with no discernable group. Just as we notice similar relationships or features, we also notice dissimilar ones, which form conceptual groups of their own based on abstraction or disparity.

Within larger visual compositions, we may notice how smaller patterns comprise a singular larger, more complex whole image. For example, equally spaced similar shapes repeated along the borders of a

Figure 3.1. Example of pattern matching by color and shape. When presented with an assortment of objects or shapes, our cognition processes them as classifications or patterns, which supports general comprehension. *Source*: Created by the author.

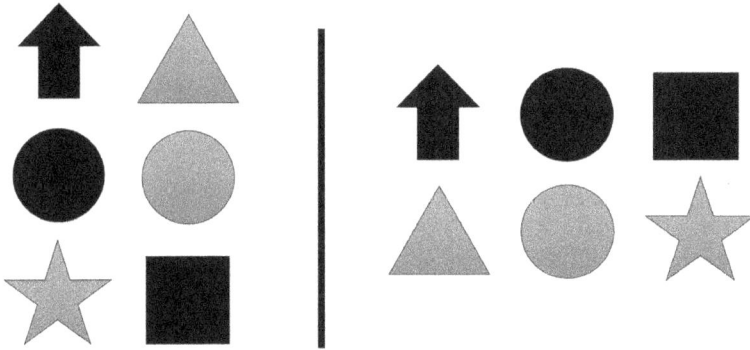

page or screen, such as flowers, may suggest a garden-themed border for a page or screen. A scattered arrangement of star shapes across the top of a page, which vary in both placement and size may suggest a night sky, particularly if there are other elements present, such as a dark background, grey clouds, and crescent moon shape. While we independently assign meaning to each individual shape (stars, sky, clouds, moon) and groups (collection of stars, collection of clouds), we can also recognize and form an understanding of the whole (night sky or landscape). Accordingly, our ability to recognize patterns from the various codes and objects present supports our comprehension of a unified whole (Barry, 1997). Pattern matching supports our higher cognitive processes, such as comparison and contrast, in which we apply and analyze information to support comprehension (Bloom et al., 1956). The cognitive acts of comparison and contrast support pattern recognition, involving our comprehension of both similarities and differences that exist between two or more objects in the same visual environment. Pattern recognition can also be seen as an extension of the Gestalt principle of similarity, where like objects are perceived as being in a similar category, while disparate ones are perceived as dissonant (Kohler, 1947). In this sense, our perceptual acts support our cognitive ones in comprehending objects and their environments.

Filtering and sorting are cognitive acts that help us focus, organize, prioritize, and sometimes discount or reject individual elements from a conceptual whole, to better understand its individual components and distinctive features. The act of filtering involves sifting new information through patterns, impressions, and meanings, which have been gathered from our collective information experiences. Consequently, our ability to filter information is based heavily on what we have previously encoded or learned from other engagements. While our perception guides our attention and initial impressions, the cognitive processes of filtering and sorting ultimately determines whether a new experience is worthy of our cognitive attention (Reed, 2022). Filtering and sorting also supports pattern matching, helping us apply knowledge from previous experiences in the analysis and comprehension of new information encountered. For example, when looking at a cluster of different colored marbles in a glass dish, we may recognize them as a coherent whole, but also notice patterns or disparity in individual marbles. We may even filter and sort individual marbles based on their brilliance, color saturation, opacity, shape, or other distinguishing features. When looking for a specific type of marble, such as color, we may temporarily filter out other characteristics to help us more effectively filter and sort from the available choices.

The act of sorting also supports the cognitive task of categorization, whereby we order and classify objects based on conceptual groups with similar characteristics, relationships, or semantics (Reed, 2022). Categorizing involves analyzing the characteristics of objects to determine their unique classes, which sometimes leads to synthesizing new categories that support our understanding. We categorize objects based on many factors, such as proximity (spatial distance), similarity (degree of likeness), and semantics (relational meaning) to help us form concepts and classify elements in our visual field (Koffka, 1935). For example, a cluster of icons in the top right-hand corner of a web page may suggest a collective function, such as a navigation toolbar. Within that context, we may classify known and unknown symbols and organize them into useful semantic categories, such their individual functions representing system settings, email, search, help, and something unknown (see fig. 3.2). Our cognitive processing involves filtering and sorting icons in the toolbar, based on function or meaning within the information environment present. Experienced users may recognize the cog-shaped icon as something that can be used to change system settings, while an envelope symbol may suggest a messaging function. Although the hourglass symbol may appear slightly different from previous experiences, it may suggest a searching feature, while a question mark may suggest system help or a knowledge base. Without additional context, the lion icon may represent an abstract or unknown feature,

Figure 3.2. Example of icons used in a navigation toolbar. When icons, shapes, or symbols are used to represent navigational tools in an information product, such as a website, our cognitive processing may involve classifying, filtering, pattern matching, and sorting these objects with prior experiences to discern function. *Source*: Created by the author.

function, or meaning within the toolbar. While an unknown shape may confuse us within a particular context, pairing it with a textual descriptor may help our cognitive processing of its specific function or meaning. This formation of conceptual categories supports our understanding of independent functions as well as how objects relate, or even support an understanding of other complex meanings or relationships in this information environment. Therefore, the acts of filtering and sorting support our understanding of categories and classes of objects by comparing them to familiar experiences and helping us form new conceptual meanings from what we experience.

To further illustrate cognitive processing as a set of collective and iterative tasks, take the example of a child assembling a Lego building blocks set. As the box is opened and upended, spilling the various pieces and instructions on the ground, the cognitive processing of the contents begins almost instantaneously. The pieces can be sorted based on color, function, shape, size, use, or other features. Initially, individual pieces may be sorted to determine if there are any missing items, which may impede performance. Next, smaller pieces of similar color, shape, and size may be sorted based on their use in performing individual steps provided in the instructions. Additionally, different sorted piles of blocks may be categorized based on different parts of the set that must be built separately, and then combined to complete the set. Regardless of the decisions of how the blocks are arranged, our cognitive processes work symbiotically, iteratively, and at varying levels of complexity, to help us successfully complete assembly of the set. Throughout the task, our perceptual processes work closely alongside our cognitive ones, which is the essence of our visual thinking (Arnheim, 1997). Through the collective processes of filtering, grouping, pattern matching, and sorting items, we are able to discern the distinguishing characteristics and conceptual classifications that form the basis of our concrete comprehension and overall information experience.

Information Complexity and Cognitive Processing

Our cognitive processes help us comprehend the unique characteristics, features, and semantics of information environments, operating at different levels of attention and complexity. While some cognitive tasks may seem simple, such as recognizing a shape, others may be more difficult, such as analyzing differences between objects, like images and tools in an information environment, with widely varied characteristics and functions. As such, our cognitive processes operate on more than a single level, which can accommodate different methods of information processing, depending on interest, motivation, need, or task requirements. Benjamin Bloom's (Bloom et al., 1956) taxonomy of educational objectives explains these differences in cognitive complexity, through a six-tiered framework describing how cognition and learning function, from basic knowledge acquisition to creating and critically evaluating new ideas (see fig. 3.3). Bloom's taxonomy is both cognitive and learning-based in application and scope, including six successive cognitive levels, which are knowledge, comprehension, application, analysis, synthesis, and evaluation. While, subsequent modifications to this model have adjusted the arrangement and naming of specific terms and levels within the model, the original taxonomy remains the more often cited and used model in research studies and applications, including its subsequent discussion in this book.

Bloom's taxonomy has also been adapted and applied in a wide range of fields, including cognitive psychology, education, instructional design, and professional certification, most notably. As a result, a wide range of educational courses, curricula, materials, and products have been developed and influenced by his taxonomy and collected works on the subject. As an example, professional certification programs have adopted the use of Bloom's model to develop certification schema, exams, and syllabi, which are organized around different and increasing levels of cognitive complexity and learning. One such organization, APM Group International (APMG), offers over 80 independent certifications in a broad range of industries such as aerospace, business, cybersecurity, information technology support, project management, and technical communication

Figure 3.3. Bloom's taxonomy of educational objectives. This multilevel taxonomy represents a series of cognitive processes, from lower to higher level thinking skills, based on increasing cognitive complexity. *Source*: Created by Derek Ross. Used with permission.

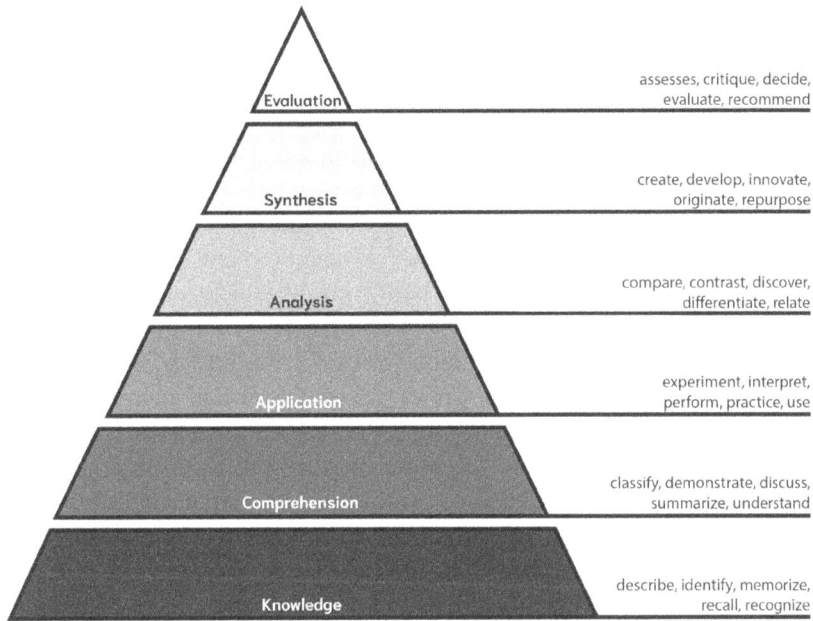

(APM Group International, 2024). In technical communication, the Society for Technical Communication's (STC) Certified Professional Technical Communicator (CPTC) certification program integrates this taxonomy as a framework to structure its three levels of mastery in professional certification, which includes foundation, practitioner, and expert (STC, 2024). Each level of mastery within the CPTC certification schema covers nine core competencies of technical communication, including project planning, project analysis, content development, organizational design, written communication, visual communication, reviewing and editing, content management, and production and delivery (STC, 2024). Each level of certification integrates the core competencies at different levels of cognitive complexity from Bloom's taxonomy. The foundation level of mastery covers basic knowledge and comprehension of the core competencies, such as basic concepts, practices, processes, and terminology. The practitioner level of mastery focuses on application and analysis of the nine core competencies, requiring candidates to identify best practices and strategies for use within specific case studies. And the expert level of mastery focuses on synthesis and evaluation, which requires candidates to create original work and defend its use of the core competencies at an expert level (STC, 2024). This taxonomy has educational applications and extensions well beyond the professional certification realm, including the creation and development of education programs, training and testing, and other information products, which span the full range of both academic and industry sectors.

An important underpinning of Bloom's taxonomy suggests the interdependence and successive nature of cognition and learning. Within the model, each cognitive level is built, and dependent upon, the previous one. The first cognitive level, basic mastery of knowledge, such as basic definitions and facts common to a field of study, is required as a foundation for the second level of cognitive processing, comprehension, which involves making basic meaning from stored knowledge. Each successive level is dependent on cognitive mastery and tasks from previous levels. For example, the ability to synthesize information, the fifth level, requires successful cognitive processing from the lower levels of knowledge, comprehension, application, and analysis. To be able to create new knowledge, we must first acquire basic knowledge and comprehension of a concept, and subsequently, develop our mastery through various practice activities, such as applying and analyzing what we learn to support such creative endeavors, such as the creation and synthesis of new concepts. Consequently, the simpler cognitive processes of memorizing and understanding basic facts and processes, support more complex ones, through practice, application, and analysis of what we know, or even the creation and evaluation of new knowledge. While each successive level depends on the previous one, it is important to understand how each complexity level functions independently in supporting our cognitive and learning processes.

Knowledge is the first and most basic cognitive level, which involves the ability to recall and recognize factual information, including memorization of simple concepts, definitions, sequences, and terms (Bloom et al., 1956). In technical communication, this might include being able to memorize and identify critical steps of the writing process, or definitions of basic terms such as information architecture or content strategy. Basic knowledge acquisition and retention serves as the foundation of other more complex cognitive functions, such as comprehension of basic concepts, facts, or processes. Acquiring basic conceptual knowledge leads to an understanding of how a specific concept functions in actual working situations. For example, memorizing a list of functions or tools used in an online help system, as a first step, eventually leads to understanding how they can be used to perform tasks or more complex operations within the help system, such as accessing troubleshooting articles or performing keyword searches. However, before this understanding of specific use is possible, we may require a basic knowledge of the range of functions and tools available, before learning how each can be used in actual working practice.

Comprehension, the second level, facilitates how we form an understanding from basic knowledge within situational contexts, including the ability to summarize, demonstrate, or discuss concepts, functions, and simple processes (Bloom et al., 1956). Comprehension is built upon foundational knowledge, which is acquired from initial studying of new information. As such, comprehension leads to an understanding of more complex relationships and uses for basic concepts, definitions, processes, and terms from our knowledge acquisition. For example, the core competency of content development from the professional certification for technical communication requires candidates to learn the definition of research methodology, followed by an understanding of the steps involved in developing one. This process may include identifying a problem, selecting data collection methods, and aligning those methods with research questions (Johnson-Sheehan, 2024). The basic knowledge of what a research methodology entails supports our comprehension of how each step functions and relates to others in the process. As a result, this comprehension can lead to performing more complex tasks such as how to evaluate and revise a research methodology for other projects with different variables. In the same that way comprehension is supported by basic knowledge acquisition, it also can support more complex cognitive tasks, such as applying what we have learned or analyzing new information we acquire.

Application, the third cognitive level, focuses on how we apply specific knowledge and comprehension to help us solve problems through design, development, or active experimentation (Bloom et al., 1956). Application demonstrates how to use practical knowledge and understanding within specific contexts and task-based situations. For example, when we learn the range of information graphics used in displaying research data within an analytical report, such as a feasibility study, and successfully master the differences between graphic types, such as bar charts, line graphs, pie chart, and tables, we can apply this knowledge to help us decide the best or most appropriate uses of each type. Depending on the type of data available, we may select a line graph to illustrate productivity trends over the course of a project, while we may select a pie chart to display percentages of how money or time were spent on individual productivity tasks. As another example, basic knowledge transfer, such as applying what we learn from one online website shopping experience to another, also demonstrates application. The basic tasks of searching and selecting items, depositing them in a virtual shopping cart, and completing the checkout and payment process, can help us learn how to perform similar sequences and tasks in other online shopping experiences. Application also supports higher cognitive processing levels, including our abilities to analyze, create, and evaluate new information products and environments.

Analysis, the fourth level, includes more complex cognitive processing, including the ability to recognize patterns, relationships, and trends in information acquired and experienced (Bloom et al., 1956). Analysis often requires a closer examination of characteristics and features within information environments, which help us discover causal and correlative relationships and semantics within those environments. For example, our analysis of multiple data points and trendlines in an infographic depicting website usage statistics may help recognize patterns of similarity and difference within the data displayed, such as bounce rates and page views for specific pages. In turn, the analysis of trends within the graphic may lead to new knowledge or a more complex understanding of the data, not obvious at first glance, such as which pages have sustained retention rates over time. Furthermore, an in-depth analysis may even help us discern specific deficiencies and dependencies, which can help us devise new strategies or innovative approaches not previously considered, such as which pages should be eliminated due to low traffic or interest. This type of complex analysis also demonstrates active visual thinking, as we explore and interpret complex data problems through focus, fixation, discernment, and concept formation (Arnheim, 1997). Consequently, analytical processes support the formation and evaluation of new ideas (or conceptual wholes), supported by the higher cognitive levels of synthesis (creation) and evaluation (assessment).

Synthesis, the fifth cognitive level, involves the ability to create new concepts, ideas, or knowledge from previously mastered ones (Bloom et al., 1956). Active synthesis is built upon the four previous levels, whereby we have acquired knowledge and comprehension of new information and actively experimented through several iterations of application and analysis. Synthesis, as the essence of creativity, supports the development of original work and new approaches to solve a problem. Much like a conceptual puzzle of sorts, which may have different outcomes and solutions, synthesis involves integrating pieces of knowledge from one complete whole and reusing or repurposing them in new and different ways. For example, when using an artificial intelligence image generation tool, we may have problems with the results from a specific set of keywords or prompts used to generate the results. After a few experimental attempts with different prompts, we may learn how to develop more efficient keyword sets to generate more accurate results. After we've become more experienced with using the tool, we may even develop procedural workarounds, which enable us to perform these tasks more efficiently. Or we might even use the image generation results as inspirations for us to create our own original adaptations for use. Synthesis is the basis of innovation, shaping our process of invention of new approaches, design concepts, processes, strategies, tactics, or even workarounds. Within professional workplaces, innovation is the pinnacle of process-mature organizations, often shaped by the creativity of their top thinkers and thought leaders. While more basic cognitive levels help us learn how to imitate or memorize knowledge, the higher cognitive levels support our ability to adapt, create, and evaluate new knowledge and strategies from what we have learned.

Evaluation, the sixth and highest cognitive level in the taxonomy, involves the ability to assess approaches, information, methods, and theories systematically and objectively based on prior knowledge and experience (Bloom et al., 1956). While synthesis allows us to create new information from learned experiences, evaluation helps us assess information in terms of strengths and weaknesses, which can also lead to improved and refined approaches to information products and experiences. Evaluation also supports the ability to make informed decisions or recommendations, based on systematic assessments, which are supported by our cognitive processing of information at each of the levels. Consequently, the ability to objectively evaluate options is dependent upon the previous cognitive levels, including prior knowledge, comprehension, application, analysis, and synthesis to formulate sound comparisons and judgements. Whether we're evaluating our own work or the work of others, this cognitive level is essential in both improvement and innovation of such work.

As an example, creating a new design theme for a website requires basic knowledge and comprehension of how to use various content libraries, design principles, development tools, and layout techniques. After mastering these basic elements, we may apply this knowledge through iterative practice and analyze various templates and tools to determine which ones are best suited for creating the design theme for the project. Once we've developed a design prototype, we can assess the features, methods, and techniques used to determine the stronger and weaker aspects of the design that may inform how to improve its presentation. And as we devise and discover new and optimized techniques, our evaluative processes continue to work iteratively with our development work to refine and improve the design. Our critical evaluation efforts may lead to additional creative experimentation or serendipitous discoveries along the way, which we had not previously considered. And as such, synthesis and evaluation often support one another in the creative process of development and learning new approaches and methods.

Bloom's taxonomy has several implications about how cognition and learning contribute to information experience. While our cognitive processing of new information may initially involve basic knowledge and understanding, our processing may move to higher cognitive levels based on our attention, interest, and motivation. As an example, we must first acquire a basic understanding of simple buttons on a device, such as a game controller or tablet, before we can optimize our mastery and performance through active experimentation and practice. After several iterations, we may learn ways to optimize our performance through workarounds, which require higher levels of cognition, such as application and analysis, to determine these performance enhancements within an information environment. When learning to play a musical instrument, such as the piano, basic lessons may involve a series of imitating, repeating, and applying basic skills through drill and practice. However, through subsequent effort and practice, our learned experience changes, eventually engaging higher cognitive levels, which help us create (or synthesize) our own musical compositions and evaluate (or assess) our own work and that of other musicians. Our cognition, learning, and subsequent information experiences are processed through different (and multiple) acts and complexity levels. As a result, our interactions with information environments, whether physical (learning to play the piano), hybrid (learning to play a computer game), or virtual (mastering the functions of a simulated environment), demand different cognitive feats to successfully process these experiences and learn new skills.

Sometimes the cognitive complexity of an information environment can overwhelm us. This cognitive overload describes the problem of receiving too much information, whether through volume, multiplicity of channels, modalities, or even excess noise present in an information environment (Baehr & Schaller, 2010). When presented with too many channels, for example, our cognitive attention may begin to filter out specific information so we can focus on the elements that are most relevant or useful (Reed, 2022). We may also restrict our cognitive processes to simpler ones, or even to a single process to focus our attention. Our selective filtering processes may select which channels we prefer to monitor, because they are expected, relevant, simpler, or better understood. When presented with more information than we can handle, our cognitive performance may also be diminished—as selective filtering and sorting may cause us to miss critical information presented because we are unable to process the complexity of the information stream. As an example, a learned, adaptive response, which requires training and practice is the ability to acutely focus on a single voice in a crowded, noisy room. While this technique is not innate, it is something that can be learned over continual experience and careful tuning of the senses. Just as we learn to isolate different communicative channels or information, our cognition can help us adapt and adjust to meet the demands of this kind of information experience, while maintaining our own preferences and comforts with the flow of information.

The Relationship between Cognition and Learning

Despite Bloom's original intentions for the taxonomy, education scholars (and others) appropriated his work for learning contexts, underscoring the important connection between our cognitive and learning processes. While any new information is processed through an iterative exchange of perceptual and cognitive processes, this eventually results in the encoding, storage, retrieval, and eventual learning of new knowledge, skills, and abilities. Our cognitive experience is also a learning one, which is the result of these actions and processes. Kolb (1976, 1984) describes our experiential learning process as a cycle of discrete tasks, which include active experimentation, concrete experience, reflective observation, and abstract conceptualization. Our cognitive filters evolve from these acts, as well as other cultural, emotional, semantic, and social aspects of our learned experiences. For example, our cognitive pattern matching helps us recognize new apps that appear on our mobile devices, based on previous related experiences. If we encounter significant differences, however, we may establish new patterns that best fit our experiences with how to launch or use newer ones. Therefore, learning can be considered a function of both perception and cognition, and our actions are influenced by our learned experiences.

Much like our perceptual and cognitive processes, learning occurs in a series of iterative cycles, through which we attempt to master or comprehend a particular information experience. Kolb's (1976, 1984) cycle of learning is a four-step integrated process, which includes concrete experience (feeling or sensing), reflective observation (watching or examining), abstract conceptualization (active thinking), and active experimenting (acting or doing), which governs how we process new knowledge or information. Our concrete experience is determined holistically by our perceptual processes during an actual experience—we sense and form initial impressions form the environment or experience. Reflective observation, suggests both perceptual and cognitive acts, including the processes of active observation and examination of the information environment, forming initial impressions while we reflect on it to learn more about specific characteristics and features present. Our abstract conceptualization involves active thinking, whereby we attempt to form new concepts or ideas from what we have experienced and observed, processing these new experiences through various cognitive and experiential filters. And finally, active experimenting involves testing our new assumptions, concepts, and ideas with others within our learned experience and in the active environment itself. After testing our assumptions, we may repeat the cycle to collect new concrete experiences or information, and process it through the various stages to facilitate our learning of new information configurations and environments.

As an extended example of how our cognitive learning occurs in this cycle, consider how an artist might study the techniques used in a watercolor image of a turtle floating in the ocean (fig. 3.4). The initial concrete experience involves feeling and sensing the visual environment. The turtle would appear to have light reflecting off its shell, which emphasizes the light and dark colors of its head, body, flippers, and tail. The same color variations in the ocean water suggest lighter and darker regions, which would seem to simulate both depth and density. Several darker shaded

Figure 3.4. Image of a turtle floating in turbulent waters. When examining a visual composition or environment, our processing and learning occurs in an iterative cycle, as we examine various features, layers, patterns, and techniques. *Source*: Liz Pohland, *Turtle Floating* [Watercolor], 2022. Used with permission.

regions on the bottom of the image may suggest surface, such as the ocean floor, while darker spots throughout the image may suggest debris or other smaller particles. Depending on individual experience, favorable impressions of marine life, oceans, or beaches might create a tone or feeling of equanimity or serenity. Reflective observation may include our impressions about action or movement in the environment. For example, the position of the turtle and its appendages may also suggest motion, such as swimming or floating. Similarly, the variations in colors in the ocean might suggest the movement of light through water, or mild turbulence. Closer reflection of the darker spots in the upper left-hand corner may suggest a floating motion of debris and other seafaring organisms. Color contrast techniques used in the image, which is rendered in greyscale colors, may even suggest full-color equivalents, based on known or expected color values. We expect ocean water to use shades of blue and green, lighting may include shades of yellow or gold, and turtles may include shades of green and yellow. In terms of artistic technique, further examination may suggest that color blending techniques such as mixing paint with water on the canvas surface with a brush or spray bottle might be a technique used to create the lighting and depth effects. Our abstract conceptualization might involve other experiences in painting or studying similar images and techniques with watercolor art. Spotting techniques could be achieved by flicking darker pigments, sand, or salt particles onto the surface of the canvas. Other blending techniques which might achieve both color variation and spotting might include spreading salt on partially dry pigments and smearing them with a brush or even tilting the paper to create the gradient effect used for the ocean water background. Finally, active experimentation might involve taking a blank canvas and testing some of the observed techniques to achieve the same effects of blending, contrast, depth, gradient, lighting present in the original image. Some creative experimentation with both subject and techniques might even lead to new methods of using watercolors with other materials. Individual independent learning is largely a function of both perception and cognition, as shown in this example, which underscores how these processes work together in creating an active learning experience.

As another example of how this cycle functions, consider how we learn to use an online application for developing presentations, which is accessed using a web browser. While we may have experience with software-based presentation programs installed on a personal computer, the online application offers added convenience and collaboration options for developing presentations. Once the online application is launched, the concrete learning experience begins almost immediately. As various elements appear on the loading screen, our perceptual and cognitive abilities begin processing the various functions, symbols, tools, windows, and terminology used in the application. These items may be classified into categories, such as familiar or unfamiliar symbols, useful or superfluous features, and new functions or processes to learn. After our initial scan of the environment, we begin the stage of critical reflection by examining specific elements within these categories to determine their particular functions, meanings, and usefulness. We may even compare familiar shapes, symbols, or tools to those found in our software presentation programs, which may help in transferring knowledge to the new workspace. Our critical reflection might also involve observing how different tools are grouped or organized in the workspace, such as design templates, text styling, or editing tools. Next, the abstract conceptualization stage might involve forming specific concepts or ideas from studying the information environment and its components. This stage might involve more detailed comparisons between features we've previously encountered in other presentation software programs and those found in the online application. For example, we might observe subtle differences in the processes of loading and saving presentations, what file formats can be imported or exported, how to switch between design templates, or how image libraries organize graphics into specific categories for ease of use. Finally, our active experimentation may involve practicing and performing tasks in the program environment. This might include creating a new presentation, selecting and customizing a design template, importing sample content, saving a presentation in an alternate format, adding audio commentary to a sample presentation, and other relevant tasks. Throughout this stage, we may test various functions to see how they differ from similar ones in other presentation programs. We may also practice using program features, which are new or unfamiliar to us, and experiment until we master their functionality and use. And finally, since learning is an iterative cycle, our active learning may involve repeating these activities to help us learn more about the application, until we lose interest or move on to another task. This continual, iterative process of discovery and refinement, while driven by our own individual interests and motivations, illustrates how both perception and cognition support our active learning, through multiple iterative cycles to improve our comprehension and mastery.

In these four stages of learning, Kolb's model also underscores the important connection between cognition and individual learning styles or preferences, such as accommodating (practical experimenting), diverging (imaginative solution making), converging (concrete problem solving), and assimilating (logical observation), which combine the acts of thinking, feeling, conceiving, and doing in various combinations (McLeod, 2017). Individual learning styles or preferences may vary based on these factors, which includes how we process information, whether it be abstractly, concretely, emotionally, or logically. Even though individuals may have one or more preferred (or primary) learning style, they may shift based on these factors, or others, such as the environment, medium, or performance task. In many assessment models of learning styles and preferences, individual learners can demonstrate some aptitude in multiple styles (such as logical, mathematical, musical, and visual-kinesthetic), rather than proficiency in only one style (Fleming & Mills, 1992; Lee & Owens, 2004; Gardner, 2008). For example, when assembling a new bookshelf from a set of visual-based instructions, we may rely more on learning styles and aptitudes that support thinking, doing, and visualizing. Conversely, when following a written process for registering for a vaccination, we may rely on learning styles that are more linguistic or logical. In fact, the environment, motivation, situation, task, and other factors may influence which learning styles and aptitudes best suit our unique information experience.

Cognitive Learning Modalities and Experience

A classic instructional learning model that focuses on sensory learning and cognitive motor skills is the VARK model, which stands for visual, auditory, reading/writing, and kinesthetic (Fleming & Mills, 1992; VARK Learn Limited, n.d.). These four sensory modalities suggest a close relationship between our perception (what we sense and our initial interpretations) and our preferences (or dominant habits) in how we cognitively process and learn new information in various modes. These learning modalities also help us understand the different ways in which information can be coded and processed, based on individual aptitudes and learning preferences.

The visual mode includes all visual forms of content, including information graphics, illustrations, diagrams, shapes, designs, space, symbols, styles, and other characteristics that incorporate primarily visual codes (Fleming & Mills, 1992; VARK Learn Limited, n.d.). Visual learning modalities (and learners) demonstrate a preference for illustrated charts, diagrams, and graphics when learning a new subject or task. Visual learners might even take notes in visual form, through diagramming, doodling, or illustrating to help encode and retain new information as part of their cognitive learning processes. Drawing visual representations of new concepts or information is a way of taking a cognitive snapshot that can be encoded, retained, and unpacked (or decoded) once mastered. Visual learners may also be particularly adept at recognizing visual symbols or images in complex patterns or environments more quickly than textual information.

The auditory mode includes most discursive forms, including voice, text, chat, email, music, recordings, and sound (Fleming & Mills, 1992; VARK Learn Limited, n.d.). Auditory learners may prefer reading or speaking aloud to learn new subjects or tasks, or even talking themselves through new ideas and concepts to process, store, and retain new information. They may develop short alliterative patterns, songs, sayings, poems, or passages that when read aloud, help stimulate cognition, learning, and mastery of new information. Auditory learners tend to be good listeners, whether to a single sound or voice, or from multiple sources. Consequently, they are also highly proficient in filtering out multiple sources of noise to focus on a single sound or voice. Auditory learning also demonstrates a preference for recorded information, which may involve repeated playbacks to help master new knowledge.

The reading/writing mode includes most forms of published or printed text, such as books, descriptions, lists, manuals, reports, and other textual artifacts (Fleming & Mills, 1992; VARK Learn Limited, n.d.). This learning modality favors reading material or use of keyword searches, which facilitates the mastery of new information. The reading/writing mode also suggests an ability to memorize and recite passages, phrases, sayings, and processes, which have been read or written previously. The simple act of writing information, whether in physical or electronic form, helps reading/writing learners commit it to memory and subsequently to recall that same information on demand. Reading/writing learners may also be linguistically adept and able to discern subtle differences in writing styles and tones. This can include tasks such as spending time making detailed written notes, outlines, and descriptions that help the repetition, mastery, and learning of various types of written material.

The kinesthetic mode focuses on preferences of learning by doing, through experience, practice, trial-and-error, or physical and virtual activities that resemble concrete experiences (Fleming & Mills, 1992, VARK Learn Limited, n.d.). Kinesthetic learning emphasizes the importance of experience and experimentation with objects and processes, and learning through the experience of doing something repeatedly, or in different ways. The act of performing a task often elicits or mirrors a sort of muscle memory, whereby repeating the physical act of performing certain tasks can be recalled and repeated, almost instinctively once mastered. As such, any physical or virtual activity resembling physical realism activity, such as practiced or repeated actions and trial-and-error, demonstrate basic kinesthetic learning.

Collectively, our sensory learning modalities vary in both the level of aptitude and preference, depending on the individual learner. We may have a primary, or dominant, mode which describes our default preference for processing new information, and other secondary, or alternative, ones that assist. For example, primary visual learners focus on information graphics, presentation, stylistic, and spatial features, but may use their reading/writing modality, to assist in learning the supporting captions or descriptions. One example of a combination type is the visual-kinesthetic style, which integrates aptitudes for visually interactive features, commonly used in many websites. Our preferred learning modalities can also be task or situation dependent. For example, when learning to play a piece of sheet music, we may rely more heavily on auditory learning to determine if we are playing it correctly, in tune or on tempo, despite the fact our primary learning mode may differ. When assembling a music stand for our sheet music, we may switch to another learning mode, such as kinesthetic, to help with the task of assembling the stand from the instructions.

While there are limitations to how these modalities explain learning in a holistic sense, they suggest the importance of how aptitudes, preferences, tasks, and even physical limitations affect how we master new information. Other more comprehensive learning theories and models incorporate and build upon these basic learning modalities, such as body-kinesthetic, intra- and interpersonal, linguistic, logical-mathematic, musical, spatial, and others included in Gardner's (2008) theory of multiple intelligences. Our cognitive learning is dependent upon our active engagement with an information environment, through multiple modalities, as well as other complex learning processes that help us interpret and master new experiences.

Another applied approach that relates to cognitive learning and immersion, is adult learning theory. This approach suggests adult experiential learning functions similarly to that of children; however, adults filter new information through their prior intellectual experiences (Lee & Owens, 2004). For adult learners, instructional content, in particular, may be less effective if it is written at a level higher than the competency of the adult learner. Therefore, the stated learning objectives of instructional content should be in alignment with the material and at an appropriate level of comprehension for the adult learner. Since cognitive processing supports our learning, it is essential to consider the unique ways in which adults process and prefer to master new information. Four factors of adult learning that help explain their processing preferences include relevance, active learning, control over the learning environment, and nontraditional forms of learning (Lee & Owens, 2004).

Relevance suggests the importance of content and its relation to the task at hand. Adult learners expect a direct relationship between what is learned and the real-world application of that knowledge (Lee & Owens, 2004). Without appropriate relevance, adult learners may filter out information they interpret as superfluous or less important to their information needs. Providing adequate content or supplemental resources for adult learners to explore can be critically important to their desire for relevance. Adult learners also prefer more involvement in learning environments (Lee & Owens, 2004). This equates to an active, involved role in the learning environment over the passive absorption of knowledge. As a result, offering choices or interactive options within the learning environment can be helpful to them in performing critical tasks. Allowing adult learners to select different information pathways, such as those present in keyword search or navigation toolbar menus, can better support their cognition and learning. Without active opportunities to interact with content, adult learners may abandon one information resource in favor of other choices that support their need for active learning.

Adult learners also prefer some degree of control over their learning experience, which is somewhat related to their preference for active involvement. Because of differences in individual learning styles and preferences, adult learners desire some degree of flexibility in how they master new information (Lee & Owens, 2004). Learning environments can accommodate this need by providing user control options, such as the ability to customize the user interface or select different methods and tools for exploring and interacting with content. Without such options, adult learners may become easily frustrated in a more restrictive environment and seek out other more flexible resources to support their learning. Adult learners also prefer more nontraditional forms of learning, which may be individualized or self-paced to accommodate how they best process new information (Lee & Owens, 2004). Providing options for users to control the pace or progression of content, such as pausing or saving progress and reviewing or skipping content support may better support adult learners. While these approaches can support both cognition and learning for adults broadly, it is important to also consider the unique ways in which they process new information in other ways as well, such as through the learner's visual thinking.

How Cognition Supports Holistic Information Experiences

Our cognition helps us conceptualize an information experience holistically, as a continual stream of filtering and processing sensory information, whether we experience it as a physical space, on a printed page, or on an electronic screen. This visual thinking suggests how our perceptual processes are supported by our cognition to help us adapt, evaluate, and learn new information and environments (Arnheim, 1997). These processes also help us adapt to new contexts and information, creating a continual, fluid information experience, which readily changes based on a number of factors. As an example, when revisiting a favorite painting in a gallery, we may not form the exact impression or notice even the same features. Differences in how the painting is presented or rendered, through various brush strokes, ambient lighting, perspective, on different textured canvases may alter how we cognitively process (and reprocess) the same image. Even the physical and digital representations of the same image may alter our perceptual, cognitive, and semantic interpretations. While every viewing of the same painting may not drastically alter our initial impressions, the nature of cognitive processing is adaptive and supports a dynamic, changing information experience over time.

To illustrate how adaptive cognitive processing functions, consider two renderings of the same image of a parrot in figure 3.5. While the

Figure 3.5. Image of a gold bird juxtaposed with a color interverted version. When comparing images or examining newer versions of an image, our adaptive cognitive reprocessing of features may alter our initial interpretations and impressions. *Source*: Mike Fuller, *Gold Bird* [Acrylic print], Society6, http://society6.com. Used with permission.

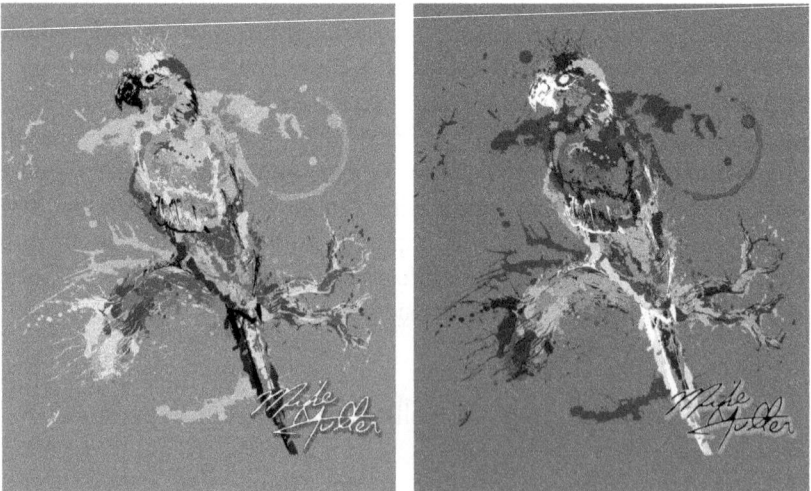

original image, entitled *Gold Bird*, is a full-color acrylic print, these two renderings are digital greyscale adaptations. The left image is a simple greyscale conversion of the original digital image, while the right image uses a negative image filter, which inverts brightness, color, and contrast. When studying the left image, we might notice its use of lighter hues to depict a brightly colored bird sitting on a branch against a mostly clear, bright sky with only a few clouds present. The subtle use of darker colors adds color, depth, and dimension to the bird, creating improved contrast against a lighter background. These techniques support a calm, serene tone and daytime environment. When studying the right image, the use of darker hues in both the bird's colors and the background suggest a different environment, such as a night sky with dark clouds and shadows. The bird would appear to be sitting on a branch, but its darker colors and glowing eye may convey a more sinister or spooky tone. We may even attribute the bird to be more predatory, due to the darker sky and overall setting. While these two images appear to be rendered from the same original image and have similar characteristics, our cognitive reprocessing of these different versions of the same image creates distinctly different impressions and overall information experiences.

Visual thinking also helps us form concepts and conceptual wholes from objects and environments, from both individual and collective char-acteristics (Arnheim, 1997). Our innate desire to form these gestalts, or conceptual wholes, includes the simple and complex semantic relation-ships between elements present. For example, when looking at a meme, composed of an image (static or moving) and a corresponding brief tex-tual message, our perceptive and cognitive processes evaluate its unique aspects, such as colors, image, movement, placement, symbology, textual message, and other factors. We may also examine the semantic meanings and relationships between image and text, foreground and background, juxtaposition or proximation of elements, and other aspects present. Our semantic interpretation extends beyond these singular aspects, which support our understanding of the collective whole. While textual codes support linguistic meaning, visual thinking helps interpret the unique visual (stylistic) and spatial (positioning) codes present. Whether we're viewing a meme or interacting with a complex information environment, our semantic interpretation can be altered by elements such as style and position, as well as substance.

The cognitive profile of a visual thinker suggests several innate abili-ties that support adaptive learning processes. This includes our abilities to adapt to new environments, to interpret semantic meaning, and to master new knowledge and skills. Even basic perceptual acts as interpreting the meaning of shapes or symbols in an information environment suggests basic concept formation, which is inherently a cognitive act (Arnheim, 1997). Once we analyze, interpret, and encode information into memory, it becomes a learned experience. Accordingly, visual thinking supports our cognition, enabling us to interpret new information and environments as collective and holistic information experiences.

The Misaligned Cognitive Experience

Sometimes information products are less successful in communicating messages, creating an information experience characterized as confusing or counterintuitive to the ways in which users think and process new information. When the information product environment creates cognitive problems for users, this can negatively impact their collective information experience. There may be one or more reasons for the mismatch between the user's interpretation and the intended product messages, which are related to cognition. Some of these contributing factors may be related to information complexity, environment design, or even learning styles and preferred modalities. Despite the product (and developer's) intentions, it is important to consider how these cognitive and learning factors affect the overall information experience, to help developers optimize information products and messages.

When the mismatch between user and product are due to complexity level, the user may encounter mastery or performance issues with products. In some cases, information may be written at a level that is too complex or challenging for the user, while in others, it may be written too simplistically. Yet sometimes, the cognitive complexity mismatch can be due to the user's level of knowledge with the subject, which may be an impeding factor. For example, a new software tutorial may initially frustrate users, particularly if the material exceeds their experience level or working knowledge of the system. Unfamiliarity with the software interface or navigation tools may create complexity issues for beginning users, in particular. Users may also have problems understanding basic functions, icons, jargon, or symbols within the environment, which describe various features. Similarly, when asked to perform advanced tasks, such as applying the use of an advanced sorting tool, which is more cognitively challenging, beginners may form a strongly negative cognitive experience. Without adequate progressive enhancement and performance assistance to accommodate them, users may quickly abandon the product due to cognitive challenges and lack of support within the system. Therefore, developers should integrate regular user testing and product review cycles, to make changes in the complexity, coverage, level, organization, or other aspects to better accommodate cognitive understanding for its users.

In some cases, our cognitive problems may be with the actual design of the information environment itself. Our basic cognitive processes allow us to match patterns, compare, contrast, classify, sort, and filter new information through previous learned experiences; however, when these basic processes fail us, it can create confusion, frustration, or eventual disinterest in a particular product. Sometimes, new, novel, or even abstract stimuli present more of a challenge if they vary from our expectations or previous experiences with similar product environments. While our cognitive processes are highly adaptable, if product environments are less flexible, intuitive, or organized in ways we can easily comprehend, this may result in a negative information experience. For example, a troubleshooting guide for the latest version of a software program might incorporate familiar genre patterns, navigation tools, organizational patterns, and other elements in previous versions. Matching familiar characteristics with new and novel ones can be a useful technique that supports our cognitive processing. Furthermore, newer products can integrate performance tips, textual descriptors, textual-visual pairs, and added help features, which support comprehension of both technical content as well as the environment in which it is presented.

Sometimes, our cognitive problems with an information product may be the result of a mismatch between the product and our individual learning styles and preferences. Often, it is not the user, but rather the product itself that creates these problems due to deficiencies in the product design. Information products also function as learning environments, particularly when users are mastering new knowledge, skills, and abilities, or attempting to understand basic product characteristics and features, such as functionality, navigation, and organization. When there are product design flaws, the user experiences cognitive challenges, which often affect performance and usability. For example, basic written instructions, may challenge learners who prefer or require visual or auditory assistance to complete the task. These limitations may also present challenges to users who have specific physical or cognitive limitations, who may require alternate forms of content to aid learning and performance.

While users may eventually adapt or find acceptable workarounds to assist them, it may take them more time or create added frustration due to lack of support for a wider range of learning modalities and preferences. As such, information products can incorporate content and features that support multiple modalities and learning preferences to help overcome these cognitive challenges.

But often, our cognitive challenges may be the result of multiple problems, which creates a mismatch between the intended and perceived information product experience. While visual thinking can help us interpret new information environments, information products should incorporate techniques which support both the perceptive and cognitive processes to facilitate both learning and optimal use. Our cognitive experience with an information environment suggests we don't just read content, but rather we experience it in meaningful ways. Information experiences are learned and remembered, until they are changed by other experiences or factors, such as improvements in product design. Whether we're comprehending new information or mastering the use of an information product, it is essential to carefully the implications cognition and learning on our perceived information experiences with them.

Implications of Cognition and Information Experience

Our cognition encompasses a wide range of acts and processes, including analysis, comparison, filtering, pattern matching, semantic processing, and learning. These overlapping processes form the basis of our understanding and learning of new information from the simple to the complex. While these processes work closely with our perception to help us interact and learn in new information environments, cognition has its own unique implications to consider, which can help us understand how to create optimal information experiences for users.

Cognition incorporates multiple overlapping processes at varying levels of complexity. Whether we're attempting to compare, contrast, filter, pattern match, discern semantic meaning or relation, or other cognitive acts, these acts function iteratively to help us comprehend and learn new information. Our cognition also processes information through previously learned experiences until we divert our attention elsewhere, storing what we have learned or simply saving it for further thought and reflection.

Cognition is also adaptable to varying levels of complexity, which includes memorizing basic factual knowledge, comprehension, application, interpretative analysis, creative synthesis, and evaluation of new knowledge (Bloom et al., 1956). Cognitive processing at different levels of complexity is one way that explains how users interpret information experiences differently, depending on their own interests and information needs. Information product development should incorporate ways in which users process new information, through various acts and at different levels of complexity to best accommodate these needs.

Cognition supports visual thinking. When thinking both visually and spatially, we engage our combined perceptual and cognitive processes in an iterative exchange to interpret and learn new information experiences. When encountering a new information environment, our senses observe the distinguishing characteristics, conceptual relationships, semantics, functionality, positioning, and other unique specifications. The handoff from perception to cognition is both iterative and immediate, much like an assembly line operating in a series of continual loops and conveyor belts. Incoming information is sorted through various processes and pathways through our perceptual and cognitive filters as we attempt to construct meaning and store patterns of new information. Consequently, our visual thinking is supported by our perception, cognition, and learning functions, as we work toward forming conceptual wholes that represent our collective information experience.

Our cognitive processes help us form holistic and cohesive impressions of the world around us. When encountering new information, our cognitive goal is ultimately to make meaning from an information environment to store for later use or further analysis. While we may selectively remember distinct parts of an experience, it may not always represent the intended one. Regardless, what we do remember forms an impression closely resembling the Gestalt notion of a conceptual whole. Consequently, our information experience is a holistic formation of our active examination, interpretation, encoding, and cognitive learning. These impressions may evolve over time, through information product iterations as well as changes in our own individual experiences. As we continue to encounter new information environments and products, our conceptual understandings will continue to evolve quite naturally over time. As a result, information products must evolve in ways that support our own changing needs and expectations for information experiences.

Our cognitive experience encompasses our cognition and experiential learning. Our cognitive processes and our learning are very much intertwined, and, in fact, to a great extent, inseparable. What we learn is often the result of successful encoding and storing information in memory. These processes are also driven by motivational factors, which are sensitive to the contexts in which that information is presented. As a result, we may also encode the same information differently, depending on our vast range of individual experiences, level of interest, motivational needs, environmental distractions, learning preferences, and other factors. Our cognitive experience is also unique—while we may be able to relate our experiences with others, we will undoubtedly see distinct differences from our unique impressions. This is because our experience is uniquely shaped by our own cognitive filters, learned experiences, and motivational differences. Every learned experience is formed by and functions as a cognitive pattern through which we interpret new experiences. While these filters can be thought of as lenses, of sorts, they are not fixed lenses through which we view the world, but, ideally, subject to continual change. These filters might include our likes and dislikes, cultural and social background, learning preferences, and other factors that influence our understanding of new information. As a result, our cognitive experience constantly adapts and evolves as we filter, sort, and learn new information environments.

Our cognitive processing is influenced by our learning modalities and preferences. Our cognitive learning integrates acts and processes that help us interpret and master new information, often incorporating different modality preferences. For example, while kinesthetic learners master information through physical experimentation, visual learners may prefer analyzing visual models and codes. Similarly, auditory learners may prefer listening to vocal quality, tone, or timbre, while reading/writing learners may analyze patterns in use of language to facilitate cognition and learning. Although we may have primary and secondary learning styles or modalities, inevitably, we all have some level of aptitude in each mode (Fleming & Mills, 1992). We may even employ different modes to suit the information environment, purpose, or task before us. For example, when learning how to complete a registration process on a website, we may rely on our visual aptitude to help locate form fields, sections, or tools used to complete registration forms, and our reading/writing modality to process specific procedural instructions. As a result, this multimodality allows us to be adaptable and flexible with our cognitive and learning processes, relying on modes that best suit a specific purpose, subject, or motivation.

Our cognitive learning processes also include our aptitudes and preferences for different modalities and coding modes as well as engaging new information at differing (and often multiple) levels of cognitive complexity. Depending on the task we are performing, we may choose (or instinctively default) to different modalities to help us process new information, whether that task involves simple memorization of facts (lower level of cognitive effort) or evaluation of the feasibility of solutions to a problem (higher level of cognitive effort). For example, when learning the rules of a new board game, we begin by memorizing the basic rules, playing pieces, game board layout, turn taking, and our possible actions. These simple acts may only require low cognitive effort. As we learn more about how the game works, we may engage our higher cognitive abilities, such as application, to help make strategic decisions on taking a turn, or knowledge transfer from other similar strategy games. The real challenge begins when we begin to engage our analytical, synthesis, and evaluative cognitive abilities in helping us assess, make, and evaluate more complex actions and strategies in our game play.

Collectively, our combined cognitive and learning processes include a wide range of abilities, which allow us to master new information experiences. Cognition functions as an iterative series of tasks from the simple, such as basic knowledge retention and comprehension, to the complex, including synthesizing and evaluating new concepts and information. These processes govern what we sense, how we interpret, and eventually how we encode and learn new information. Our cognition supports our visual thinking, helping us interpret both image and space, form symbolic concepts, conceive semantic and relational aspects, solve problems, and learn from new experiences. Our collective interpretation forms the basis of our information experience, representing a conceptual, holistic understanding of an information environment and its content.

Chapter 4

Information Environment Design

An information environment encompasses the look and feel of an information experience, which includes visual, spatial, and textual codes, as well as other integral features, such as interface layouts, interactive forms, navigation tools, and visual identities. While the environment can be static, dynamic, interactive, and even hybrid, it represents the space in which content is presented to users. Within an information product, we are conscious of both content and environment, which function in a similar way as figure and ground in visual compositions. Both content and environment have communicative value in information products, helping users master content and the interface through which we interact with information products. For example, many online information products incorporate expected features such as a screen-based interface, navigation tools, visual elements, interactive media, and content. While our experience is highly influenced by what we see and interact with on the screen, our information experience can also be affected by the ways in which we customize features and settings governing both access and use. The environment itself, much like our information experience, may also change dynamically, based on these specific conditions and settings, as well as our own interactive responses.

Information environments function as the interface between information (or content) and user. The term *interface* describes the collection of content and features that comprises a product's information environment, and is also a way of seeing the whole (Johnson, 1997). This whole suggests an important connection to Gestalt theory, and our perceptual and cognitive processes, which helps us form holistic impressions of our information experiences. Within electronic information products, the interface encompasses everything we see on a screen, which includes both physical and virtual aspects, such as navigation tools, buttons, tabs, windows, and other haptic devices, like keyboards, mice, and touch screens, used to interact with content presented. While both human perception and cognition are critical to our understanding of information experiences, our expectations are often built upon conventions and practices acquired from our practice and use of these interfaces. The information environment can incorporate common and uncommon navigation or organizational patterns, which influence how we browse, interpret, navigate, read, and search for information. Therefore, understanding how information environments and their unique characteristics influence both development practices and uses is essential in supporting both content and user.

The Information Experience Triad: User, Content, and Environment

Information experiences encompass three essential elements of an information product: the user, content, and the environment. Each element interacts dynamically with the others, and each plays a critical role in the overall information experience. While this chapter focuses primarily on the environment, or interface, it is important to understand the distinction and overlapping characteristics of each, as well as those that exist between them in the holistic information product experience. While the first several chapters of this book focus on the user, it is important also to consider how users interact with and are influenced by both content and environment in the overall information product experience.

As a result of electronic technological advancements widely incorporated into today's information products, readers have transformed into users who may interact with the information environment in vastly different and unique ways. Users are adaptable, learning as they go, and building on previous experiences and skills acquired from other information experiences and products. Experienced users are capable of finding workarounds and adjusting their behaviors to fit the information product environment more readily. As a function of both perception and cognition, users are naturally curious about what is new and novel, driven by specific motivations or tasks, but often are open to diversions or different experiences, which may be relevant or useful. Through independent exploration, electronic information environments often incorporate associative hyper-links, which connects content topics by semantic characteristics, offering users related and serendipitous content choices. Users expect accuracy and reliability in content and will seek out a wide range of factors to assess its usefulness. This assessment might consider comments from users, sum-mative ratings, successful branding techniques, and other factors. Users also expect a certain degree of alacrity in their information experiences, partly since information technologies continue to accelerate the access of content, whether downloadable or streaming. Consequently, a higher level of motivation may be required for users to tolerate longer load or wait times when accessing content, while a lower tolerance is typical for less important tasks, or when users know they can seek alternate, more reliable information resources.

Users often have specific expectations with information product envi-ronments, such as flexibility in how they access and use those products. As one unique expectation, flexibility offers users choice, such as navigating or personalizing their experience, which may influence continued or repeated use of the same product, based on a perceived favorable experience. Users may also expect experiences that are organized and intuitive over ones less so, in part because these characteristics often require less cognitive effort to learn and use the information product. When users can rapidly discern the functions and organization of an information product, their performance will improve, as well as their perception of the information experience. Undoubtedly, there is a wide range of other user expectations and preferences, which may be unique to a particular product environment that developers can discover and accommodate through careful research. Accommodating these expectations can also help developers create infor-mation experiences and content, which builds lasting favor and loyalty with its information product users.

Content has its own unique distinguishing characteristics and fea-tures, which are optimally designed and written to support both the user and the information environment. Content forms can be static, or fixed in its properties, or dynamic and changing, based on specific conditions or user interactions. Content incorporates the visual, spatial, and textual codes used to communicate messages, whereby image, space, and text work seamlessly to create both simple and complex forms of content. Interactive content can be browsed and searched using a wide range of navigation tools, which can affect how content is displayed, filtered, organized, and ultimately used. Content can also suggest differences in how we perceive an information experience and its sense of realism. Simulated environments, in which enough characteristics and features suggest realism, can make virtual content environments, such as games or simulations, communicate a seemingly real information experience to users. For example, tasks that can be performed in the physical world, such as flying an airplane, can also be performed virtually, through a software environment; however, the physical and the virtual environments carry different sets of expectation with regard to the levels of perceived realism. Heim's (2000) works on virtual reality and realism describe different levels of virtual realism from basic simulation to full-body immersion, describing how interactivity and realism function differently based on the environmental characteristics in a wide range of contexts, including commercial, educational, entertainment, and even health-care-related applications. Interactive content can also be both visible, or immediately perceived, and hidden, such as in the use of scripted algorithms, which remains obscured from view until triggered by special conditions or user actions. While some content forms require our active interaction, we may prefer to passively experience. Accordingly, information products may require a complex and diverse range of content types that appeal to a wide range of uses, often forming unique hybrids. Content hybridity suggests that features are not always assigned to mutually exclusive binaries, such as static versus interactive content, or even print versus electronic content. Often, information environments include mixed combinations of content, media, modalities, and interactive properties.

Content can also include multiple overlapping layers and levels within the interface, which includes surface, structural, and code. Garrett (2010) describes a multilevel content model that integrates key features of both content and environment, such as the visual, navigational, structural, and interactive properties of an information product. Many of the surface level characteristics focus on the textual, visual, and spatial codes, which are visible in the product interface. This content includes the overall messages present in the information product—its sections, paragraphs, pages, information graphics, and spatial grids used. Structural level content characteristics include the organizational patterns used to arrange pages into linear, hierarchical, and hypertextual configurations throughout an information product. Structured content can take on many forms, which are both visible and hidden from the user, such as site maps, navigation menus, heading levels, and various content tagging methods used. These elements help organize and communicate the structure and semantics of content, helping users understanding arrangement, findability, and usefulness of content present. Code level content is rarely visible or present in the interface, but, rather, operates behind the scenes, incorporating the use of markup, scripting, and programming languages. These languages facilitate the presentation, structure, organization, style, position, and interactive properties of content. Together, layers of surface, structural, and coding level content types function together holistically to communicate various messages and their semantic properties to users of information products. The relationship between content and the information environment is symbiotic and somewhat similar to the Gestalt notion of figure/ground, where individual content units function as figures, presented in the foreground, and the interface functions as the conceptual ground. For example, in a website, the background of any page includes elements present in the interface, such as the overall spatial grid layout, the overall design scheme, browser controls, and so forth. The content presented on individual pages, such as textual descriptions, headings, titles, information graphics, and other elements, are featured in the foreground for users to read and use for various purposes. While each of these elements has its own unique characteristics that set the stage, so to speak, together, they convey a singular, holistic message combining figure and ground elements into a complete presentation.

The information environment itself, or interface, encompasses both physical and virtual characteristics, which operate the interaction between user and content. Physical aspects may include the use of cameras, keyboards, mice, microphones, screens, touchpads, and other haptic devices. Virtual aspects may include buttons, images, hyperlinks, menus, media controls, navigation tools, and other content elements present in the information space, which allows content interaction and use within an electronic environment, such as a computer or mobile device. The relationship between the user and the interface is also an interesting and symbiotic one. While there is a clear physical separation between the two, when considering the virtual environment, this separation becomes less distinct. Johnson (1997) suggests the interface functions as an extension of the human user, who reaches out into virtual space through physical devices (computer, keyboard, mouse, screen, etc.) and virtual controls (hyperlinks, search forms, toolbar menus, etc.), which enables the user to access, browse, filter, search, and sort information. Within more complex information environments, human users can create their own virtual avatars and profiles to establish their presence and interact in more complex ways. As such, users are often afforded some degree of control over both their interactions and presence within such information environments.

Despite the presence of physical characteristics, information environments are largely virtual for the user, particularly in electronic information products, whereby users extrapolate characteristics of a physical experience, which is simulated on a screen. For example, in a three-dimensional virtual walkthrough of a living space, such as a virtual home tour, although the actual presentation is a series of interactive tools and seamless virtual images, this experience can be perceived as actually exploring and walking inside a physical space, if the realistic properties are sufficient to engender such an information experience. Specific physical, spatial, and sensory characteristics can also create mental and emotional impressions within users, comparable to real experiences. Our thresholds between realism and superficiality, may also vary depending on these factors, as well our own unique desires, motivations, and prior experiences stored in memory. Customizable content options can create information experiences that are rich, interactive, dynamic, and usable to fit a wide range of functions and purposes. Information environments are increasingly complex, particularly in hybrid and electronic information products, encompassing all aspects of how users, interfaces, and content come together to form the information experience.

The Design and Structure of Information Environments

In a comprehensive study on research topics in technical, business, and professional communication, Carradini (2020) identifies concepts most frequently and specifically associated with technical communication, which include knowledge, data, framework, usability, and user. These concepts closely relate to the components of information experience, and, in particular, information environment design. Within an information environment, content (knowledge or data) can assume many forms, including interactive, numerical, spatial, textual, and visual. An information structure (framework) can vary widely based on the nature of content, context, interaction, and use. Information environments depend on successful function and usefulness (usability), often involving rigorous prototyping and testing through iterative cycles to enhance their accessibility and use. And throughout the entire development cycle, a user-centered design process emphasizes individual performance (users), helping ensure products align with their unique expectations, needs, preferences, and uses, where possible. Consequently, users themselves can be highly influential on environment design, providing feedback through comments, focus groups, interviews, ratings, surveys, and through other analytic methods. User-centered design, as a commonly used design practice in software engineering, routinely integrates these various types of user analytic research, which inform decisions made in the design and development of information products.

Information product environments have also become increasingly customized, digital, portable, and responsive in nature, to accommodate use. Information product environments have become increasingly dependent on technologies and tools, requiring an increasing breadth and depth of knowledge and skills to use them. Additionally, the information environment can be influenced by the user, authoring tools or structures, as well as the content itself. Specialized tools may be helpful in creating information environments that align with the expectation of the particular product category or family, which may include word processing, spreadsheet, presentation, instructional design, web-based platforms, social media platforms, and others. The authoring tools (or programs, platforms, etc.) themselves, as well as their embedded organizational structures (such as templates, forms, etc.), can influence the environment design as well.

For example, within a simple presentation software program, design templates are available for use, which help organize content into bulleted lists, columns, tidy information graphics, and other forms. These templates are a function of the system that influences the layout, design, and organization of content, when not modified to any great extent. While some programs provide greater customization options than others, working within standardized frameworks presents its own set of variables that influence environment design. These organizational frameworks help users quickly develop new presentations, but with the trade-off of limiting how information is chunked, designed, organized, and presented. In a sense, the system acts as an agent, influencing the overall design of an information environment, which invariably can affect how content is received, both positively and negatively.

Depending on a variety of factors, such as the purpose, types, volume, maturity, and functions necessary, content can also shape information environment design, particularly as an information product evolves and matures. Two approaches that describe how content evolves within an information product include a folksonomy, or user-driven approach, and a taxonomy, or content-driven approach (Governor et al., 2009). These approaches consider how user, content, and environment influence both the design and structure of information products in different ways.

Information folksonomies develop over time, based on user preference and consensus, to serve as the guiding mental model for developing catalogues and lists of topics, such as a knowledge base or wiki. Smith (2008) defines a folksonomy as a bottom-up classification strategy, which often emerges from user-generated content and preferences, including content tagging. Content tagging allows users to create their own descriptive keywords (or tags) and links that best fit their contributed content (Governor et Al., 2009). While this approach favors users throughout an information product's evolution, it is not without its challenges, including inconsistent uses of terminology and lack of standards in both content editing and metadata integrity. In particular, when the folksonomy includes a large group of contributing users, this can create obvious usability and findability problems over time within the product. While a folksonomy may have strong user-centered and content-centered aspects as its underpinnings, this approach tends to rely on social construction of knowledge over specific structured information taxonomies as guiding organizational patterns. It also suggests the information environment is regulated more by the user base itself than specific established content standards.

Information taxonomies, on the other hand, focus more on the content assets themselves than individual users (Smith, 2008). In some cases, a small group of subject matter experts drive decisions made in organizing material into topics and coherent information structures. These decisions might be based on several factors, including existing content patterns, metadata standards, content volume, analytics, and other mitigating factors. In more complex information products, the content can be organized by the system itself, such as in a content management system, which can accommodate and adapt more readily, as the information product evolves. Adaptive content, which adjusts, filters, and displays content within a content management system, can also create unforeseen challenges for users (Rockley & Cooper, 2012). Within adaptive content systems, taxonomies are often autogenerated, which can create potential usability problems, particularly if users fail to comprehend the system-generated information structure. While an information taxonomy may not incorporate specific user preferences, instead it focuses on designing an information system and environment that is based on analytics and content trends present in the information product.

While the information environment of a product can be developed, organized, and structured using a wide range of approaches, many often are crafted as organizational hybrids, combining content standards, different structural patterns, and user preferences. And while user aptitudes, internal processes, and preferences shape their information experiences, the various features within an information environment shapes this experience as well, through automated processes, interactive properties, organizational patterns, structures, and templates. Each of these aspects influence our perceived information experience and should be accounted for in information environment design. But equally important is what the user cannot visibly see on the screen, operating behind the scenes, which affects how interfaces behave. This may include algorithms and scripts, present in more dynamic information environments, which adapt and change based on our interactive responses and other preset conditions.

Interfaces facilitate the means by which humans and machines interact. Marshall McLuhan (2017) suggests that *the medium is the message*, which emphasizes the importance of the information environment, or interface, as a critical factor in information design, whether it conveys a series of simple messages or a complex functions and features, such as those found in software systems or websites. Interfaces embody unique characteristics, which are deliberately and purposefully integrated into information environments by their product designers. Interface characteristics can also be attributed by their users, through their perceived information experiences with product environments. The visual and spatial thinking processes that help us perceive and comprehend the increasingly complex landscape of information products and interfaces also help us find meaningful ways to use these products effectively. Accordingly, visual-spatial thinking is an articulation of our combined perceptual and cognitive abilities, facilitating our interpretation of the various visual, spatial, and textual codes present in an environment (Johnson-Sheehan & Baehr, 2001). Our holistic understanding of these codes supports basic performance tasks within these environments, such as browsing, searching, scrolling, and other interactive behaviors. Just as information designers strive to create cohesive information experiences, the visual and spatial thinking processes also strive toward a similar goal: forming a complete and conceptual whole (or holistic information experience) from the various codes and characteristics present in an information product environment. Often, our behaviors are guided by these collective visual-spatial processes, whereby information environments are observed, interpreted, and acted upon.

How Information Environments are
Informed by Hypertext Theory

Our conception of an information environment and its features, whether a single document or a complex virtual environment, is what Barry (1997) refers to as a mental gestalt, whereby we formulate a concrete understanding of what might initially be an abstract environment, with some familiar elements. The information environment itself includes characteristics that support content, such as design layout, information organization, navigation, semantics, and spatial and visual presentation. Many of these characteristics support our understanding of information environments and are derived from early research and theories dealing with electronic information products and, most notably, hypertext theory.

One of the earliest and frequently cited articles on hypertext theory was from Vannevar Bush, who in his 1945 essay published in the *Atlantic Monthly*, described an information sharing device called the Memex. This device was conceived as a mechanical-electronic hybrid information storage and retrieval device, with striking similarities to modern networked computers with Internet access (Bush, 1945). Bush's main goal for the Memex was to create an information sharing tool that allowed researchers to pool their collective works with others, with the added capabilities of organizing and notating information in the system in more semantic and interactive ways. As originally conceived, the Memex would have knobs, levers, and display screens, which permitted users to search, notate, bookmark, semantically link, and connect to other information sources within the system. As with any technological device of a particular era, in this case the 1940s, the device's conception was somewhat bound by mechanical conventions and features commonly used at the time. Future advances in information technologies eventually saw the embodied characteristics of the Memex achieve practical means well beyond his initial conception, particularly in electronic information environments and products to come. Although never fully realized or produced as a working device, the Memex inspired other inventors of information technologies for decades to come, including, most notably, Tim Berners-Lee's conception of the World Wide Web as we know it today. Berners-Lee (1999) even acknowledges these influences, including characteristics present in Bush's Memex device and Nelson's Project Xanadu, both conceived as early proto-hypertext systems.

Many technologies conceived in Bush's era were largely conceived as mechanical and electronic hybrids, with product features and interfaces dominated by physical controls, such as buttons, knobs, levers, and screens, rather than virtual characteristics that would later evolve in the computing environment. To illustrate some of the limitations of the Memex, information was stored on microfilm and displayed on monochromatic screens, controlled by buttons and levers. While the use of display screens suggested electronic characteristics, the use of microfilm suggested more print or mechanical based characteristics.

Derived from many early information systems, hypertext theory described predominantly nonlinear, associatively linked content environments for information products. As electronic information environments have evolved, terms such as reader and text have been replaced by others, such as user and content, which better align with a predominant electronic information experience. Within hypertext systems, such as websites, narratives can also be communicated through many of the same characteristics, including associative linking, hypertextual structures, navigation pathways, and predictive semantics (Johnson-Sheehan & Baehr, 2001). While older print-based information products, such as manuals and reports, were governed by more stable concepts of authorship, structure, and textual composition, electronic information products changed many of these notions, causing both author and reader to rethink their roles in content creation and consumption (Lang & Baehr, 2023). In electronic (and hybrid) information environments, content authoring can be a more highly structured and collaborative process, while supporting navigation systems can include multiple methods and tools, and content can be associatively linked and dynamic, interactive, or multimodal in nature, to fit a wider range of users and uses. For example, in content management systems and wikis, content can be created by multiple authors and users, incorporate multiple navigation tools for browsing and searching, provide multiple pathways for users to explore and research, and provide useful hyperlinks to other related information resources. As such, these hypertextual structures, as organizational patterns, are created naturally as a web of trails as new content is added to an information system (Bush, 1945). And these structures can evolve both deliberately and organically to become independent systems, such as fully developed knowledge bases or information wikis managed by a healthy user community and development team in consort.

One such knowledge base, the Society for Technical Communication's (STC) Technical Communication Body of Knowledge (TCBOK) project, mentioned in an earlier chapter, functions as a free-access, searchable, and evolving information resource for both research and practice (see fig. 4.1). The TCBOK features a complex hierarchical structure of content topics on the full range of concepts, methods, practices, and processes used widely in the discipline of technical communication. Content topics can include articles, images, links, media, and reference lists, depending on the detail and scope of individual entries. These content topics are organized into several information levels, based on their category, subcategory, and identifying keywords. Initially, the TCBOK was created as a structured fully collaborative informational wiki featuring content curated and developed by novice and expert academics, practitioners, and students in technical communication. As the TCBOK evolved into hundreds of individual content topics, both structure and functionality evolved as well, from a collaborative wiki into a fully developed knowledge base, which functions as a Wikipedia, of sorts, for technical communication topics (Hart & Baehr, 2013; TCBOK, n.d.). As a wiki, the TCBOK evolved organically as new topics were created and edited, growing in both scope and size. As a result of this growth, the TCBOK required a more sustainable information structure for the maturing information product, which included improved navigation, optimized tagging, and reorganization of existing content topics.

Figure 4.1. The Technical Communication Body of Knowledge interface. The TCBOK's information environment provides users with a variety of searching and browsing tools to explore its vast range of content topics. *Source*: "About the Technical Communication Body of Knowledge," Technical Communication Body of Knowledge, http://www.tcbok.org. Used with permission.

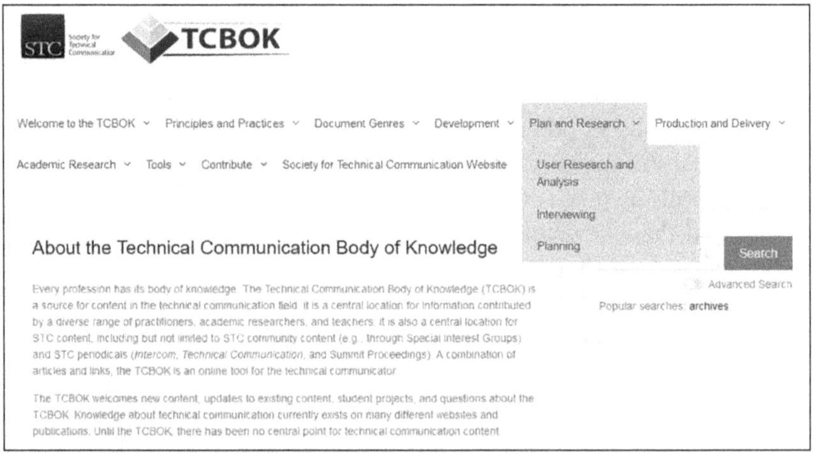

To help with its growth and evolution, an expert group was formed to strategically develop and plan revisions for the project and its growing information taxonomy. New categories, keywords, and tags were added to support the emerging complexities in the knowledge base, which improved the hyperlinking and navigation capabilities, which included linked content both hierarchically (from general to specific categories) and hypertextually (associatively based on content relatedness or semantics). Within this new hierarchical-hypertextual hybrid system, the TCBOK topics could be explored using multiple navigation tools, which support associative browsing and searching, whether by content index, keyword search, navigation toolbar, site map, or tag cloud application. These changes to the TCBOK subsequently improved its information environment, providing more flexible navigation options and a more scalable system, which facilitated better categorization and tagging for future content topics.

Many contemporary information-based wikis and knowledge bases share similar hypertextual features. While early conceptions of hypertext theory and its distinguishing characteristics were not exact blueprints, they provide theoretical foundations for today's information products and their environments, which embody many of the core features of early hypertext systems. These characteristics emphasize aspects such as collaborative authoring, content focus, hyperlinking, hypermedia, intertextuality, and multipathed structures, among many others (Baehr & Lang, 2019). While hypertext explicitly refers to the ability to link text, the simplicity of this term is underpinned by several semantic, spatial, and structural aspects of information environments. One of the core concepts of hypertext theory is that associatively (or semantically) linked text could create limitless possibilities for information structures, organizational patterns, and collaboration, creating new ways of how texts could be imagined and read by users. For example, semantic markup, used in Hypertext Markup Language (HTML), can denote both function and meaning of textual content, which may include elements such as titles, search keywords, navigation tools, alternate image descriptions, headers, footers, and others. Using semantic markup to communicate meaning of these elements supports overall user comprehension of the functions, purposes, and uses of information products. As such, the information product environments support both function and semantic messages, which add yet another layer to the overall information experience.

Another core concept of hypertext theory involves the capability for users to contribute content through collaboration and interaction with an information system. This collaborative content can be abridged, edited, moderated, metadiscursive, original, and supportive in nature. Often, content has different functions within a hypertext system—original or primary content, abridged or edited content as secondary content, or even metadiscursive content, which provides related commentary on a particular subject. Within hypertexts, content can also be reused and repurposed, allowing for communicative efficiency where writing something once and reusing it multiple times is possible within the same product or system. While content reuse has become a tenet of electronic information development, its early conceptual notions in hypertext theory are well established. Even Bush's (1945) proto-hypertextual Memex featured information capabilities that allowed content to be accessed, annotated, linked, organized, referenced, and repurposed at the will of the user. Within this system, content itself can be a product of user interaction. This aspect of hypertext theory expands the semantic landscape of content as well as the interactive and intertextual characteristics of information products, creating more dynamic information experiences for users.

While information products rely heavily on both textual and visual content creation and collaboration, they also rely almost exclusively on spatial characteristics, including the space between visual elements, connections between hyperlinked text, and even the spatial configurations present in information architectures used. The spatial characteristics of an information environment conveys important structural and semantic details that support performance and use. This notion is supported by the Gestalt principle of proximity, which suggests space can communicate information about the relationships between individual elements within an environment based on their positional characteristics (Koffka, 1935). Examples of this within an information environment might include the use of content markup, heading levels, navigation menus, site maps, and other elements that suggest relatedness between content units and functions. While information environments may typically arrange content linearly or hierarchically to communicate processes or specific taxonomies, often other spatial configurations, such as hypertextual or even customized structures, can facilitate optimal functionality and use of these product environments (see fig. 4.2). And while simpler information products may require one structural pattern or type to organize content assets, more complex ones may require multiple structural types, or even customized structures to accommodate multiple functions or features.

Figure 4.2. Examples of information environment structures. Information environments can include a wide range of structural configurations for organizing content and functions, including linear, hierarchical, hypertextual, and customized structures. *Source*: Created by the author.

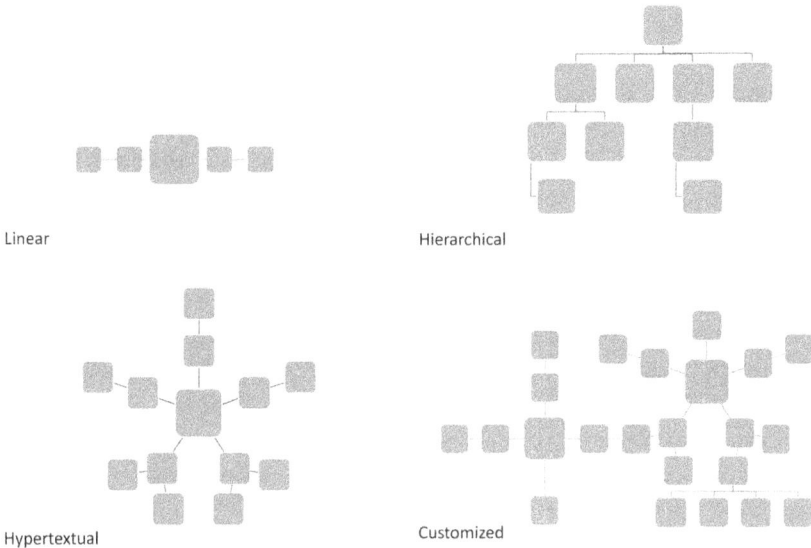

Linear

Hierarchical

Hypertextual

Customized

Linear structures are the most basic configuration, which are often used for sequenced tasks or processes that require a step-by-step progression. While linear structures offer less flexibility in terms of navigation, either back or forward in a sequence, they can ensure users follow predictable and stable patterns when navigating an information environment. Within larger information products, such as large-scale websites, linear structures may be used, in part, for specific functions, such as user registration or purchasing items. Often, these simpler functions require users to be directed toward a specified path through a sequence of steps to successfully complete a task. In this way, linear structures can support larger, more complex information structures, providing patterns for simple functional processes. In some instances, linear structures may also be preferred for their simplicity and use to optimize performance within the information product environment. In some cases, structural complexity should be reconsidered where simpler structures may offer a more efficient information experience for users.

Hierarchical structures build upon the basic concept of linear structures; however, they can accommodate more complex taxonomies, such as content organized from general to specific topic. These hierarchies can vary in terms of breadth and depth, and depending on the complexity and volume of content within an information product. Often, navigation menus and site maps communicate these hierarchies to users within the information environment itself, supporting overall comprehension of content, function, and organization. However, other techniques, such as the use of headings, nested breadcrumb links, or even visual styling (such as color, shading, or spacing) can also communicate the various information levels within an informational hierarchy. Hierarchical structures also help users understand semantic relationships between different levels and sections present in an information environment, such as content categories or related functions.

Hypertextual structures accommodate more flexible navigation and organizational options, where content can be accessed or linked based on associations. These associative (or semantic) relationships can be communicated deliberately in the information product environment, or even established by the user, through their browsing history or other interactions. Ideally, within these structures, users can access any related content unit (topic) or node, regardless of its location within the information structure, through active hyperlinks, keyword searches, or other interactive means. Through these features, hypertextual structures can provide users with choices such as *you might also like* or *other users also viewed these items* as serendipitous options related to the user's active interest. Often, these kinds of choices are provided to users when they are browsing and searching options in online shopping experiences. And when combined with linear or hierarchical patterns, hypertextual structures can provide this flexibility and multiplicity in how users interact with information environments and their contents.

Customized structures accommodate both complexity and flexibility for both the information environment and the user, but may require additional effort to ensure optimal usability. Essentially, these structures combine features of multiple structural types, and in some cases, create unique organizational patterns, which support the functions, purposes, or uses of an information product. Often, customized structures represent system-centered designs, which accommodate designer preferences, but require such user-centered design elements to support performance. While such structural complexity supports higher levels of customization, these structures can create additional complexity for users without these accommodations for both access and use. Communicating structural elements coherently to product users may require additional features within the information environment, such as the use of contextual clues or messages, guided demonstrations, help systems, performance tips, or suggested workarounds. Customized structures may also require constant monitoring, testing, and revision to ensure both accessibility and usability standards are met in future information product iterations.

Ideally, information structure should align with function to ensure the information environment supports users and the intended uses of an information product. For example, hyperlinks accommodate more complex, cross-referenced semantic relationships between individual content units within an information structure, which, in turn, supports interactive browsing and searching. Also, consider the differences between print and electronic forms of this book, or any other text. The printed version may have an index, table of contents, list of figures, and references, which organize its contents hierarchically or topically by chapter or section number, heading, keyword term, or reference. The electronic version, however, can incorporate hypertextual features, such as interactive hyperlinks that connect related sections or topics in the book, based on associative or semantic relationships. Additionally, the electronic book can incorporate advanced keyword search tools, which accommodates faster access to relevant content. Illustrations within the electronic product can include interactive properties, such as animation, multidimensionality, or rescaling or resizing options, which allow them to function as more than simple, static images. And depending on the information environment, other tools may allow users to create their own bookmarks, notations, or links between individual sections or content topics, which allow them to annotate and append the existing text. Consequently, the semantic, spatial, and structural aspects of an information environment can create a different and more dynamic information experience for users. These characteristics are also well supported as unique features of hypertext systems, which continue to find application and use within today's information product environments.

Technological Convergence and Evolving Product Environments

As discussed, Vannevar Bush's (1945) Memex incorporated a wide range of technological features into a proto-hypertextual system, which included the capabilities of information creation, notation, retrieval, sharing, sorting, and storage. Much like other technologies that followed, newer inventions and iterations would incorporate new features as well as a convergence of older ones. As it was conceived, the Memex incorporated separate established technologies, such as a microfiche reader, television screens, and other early mechanical computing devices, with several new features, into a new information sharing system. This technological convergence describes how previous individual technologies can be combined with other newer features and remediated into newer, useful, and upgraded information product. Convergent and remediated technologies retain some characteristics of their predecessors, but also evolve to include new features (Bolter & Grusin, 2003; Jenkins, 2006). For example, smartphones represent technological convergence of other useful portable technologies, including cameras, flashlights, global positioning systems, and personal computers. These convergent technologies allow users to perform a wide range of functions, which decades ago required separate technological devices. But as technological products and their features converge, so does the content of these information systems.

Convergent content is common in many component content management systems, where content units are written granularly for reuse across multiple platforms, settings, and publication formats. For example, within a web-based content management system, content chunks encompass a range of topics, functioning as modular components that can be combined into more customized and complex pages. Individual content units may include a short title, short description, keywords, hyperlinks, images, and many others. Each content component can be retrieved and reused to create a wide range of published forms. Within such a system, a keyword search on accessibility may return several individual content topics, such as accessible design, accessibility standards, and web usability, which may be presented on a single page for users. As such, content can be customized and transformed from separate content units and reassembled into a wide range of published forms, based on simple user input through keyword search. This convergence can also incorporate multiple modalities, such as different forms of media, like audio, image, text, and video content, into a single interactive presentation.

While the early World Wide Web was limited in its bandwidth and capabilities, the notion that media could be combined from multiple sources and shared across a vast information network was still an early promise. Theodor Nelson's work on Project Xanadu in the 1950s built upon the characteristics of Bush's Memex, proposing an actual working model of an information storage and retrieval networked system. Within his system, a form of hypermedia, which he called a hypergram, was an early form of interactive convergent media content (Nelson, 1992). Some early examples of hypergrams included image maps and simple animations, which later would evolve into more complex visuals such as multidimensional (or layered) images and scalable vector graphics. One prevalent and commonly used example in social media content is the meme. Memes can be static or animated, typically pairing verbal and visual messages together into a single composition to reflect a particular message, mood, or tone. As hypergrams, memes also have embedded semantic messages, which are often culturally or contextually bound, which can be used discursively to communicate deliberate messages, or metadiscursively to comment or react to specific messages.

Often as the result of technological convergences, information products and environments change and evolve over a product life cycle. As both physical and virtual characteristics of these products and environments converge into more hybrid forms, undoubtedly, the resulting information experience will continue to change for users. This technological evolution may result in an amalgamated or upgraded form of the original version, or simply the next iteration in a product's development. While some latent or supplanted features of a product or environment may be abandoned, others may be repurposed and reused to create these newer, more convergent forms that evolve (Jenkins, 2006; Baehr & Lang, 2019). For example, a personal computer offers both physical features (keyboard, mouse, monitor) and virtual features (simulated desktop space, icons, clickable hyperlinks, interactive images). These information environments may incorporate capabilities such as downloadable content, printable pages, searchable content, and virtual models for users to experience content. As computing technologies have evolved, older supporting features and peripherals, such as floppy disk drives and monochromatic display screens, have been replaced and repurposed with new features to enhance the speed and visual presentation capabilities of information systems, such as computers and other peripherals.

Technological convergence often characterizes specific shifts in dominant features or forms for newer ones. Walter Ong (1983) suggested that when newer technological forms evolve, some previously dominant features are not entirely supplanted, but could remain latent for some time, or even reemerge in later evolutions. Ong conceived of major shifts in communicative technologies as different communication eras within which technological products were situated. These included the oral, written (or manuscript), print, and electronic eras, each embodying unique characteristics of how content is created, organized, shared, and used (Ong, 1983). Within each of these communicative eras (oral, written, print, electronic), different dominant forms of communication represented new and converged technological features and forms (Heim, 1999). For example, the oral communication era included orators, speeches, and forums as their touchstones, while the written era had its scribes, manuscripts in the form of codices or scrolls, and vast libraries. Subsequently, the print era had its authors, printed texts, and published collections, while the electronic era has multiple collaborative authors, cloud storage, digital documents and media, networked information platforms, and social media (see table 4.1). While Ong's initial work on the topic of communicative eras predated much of the current landscape of electronic information product environments, his notions on these and other evolutionary shifts and emerging characteristics provide a good foundation for how we understand these notions.

As an example of how information technology from one era converges and evolves into newer forms, consider the use of the telegraph. First developed in the mid-1800s, the telegraph enabled the electronic transmission of simple short textual messages, which would later be replaced by the telephone, a talking telegraph of sorts, and later by other technologies, such as the text-enabled smartphone, which enabled the use

Table 4.1. Four Communicative Eras and Key Features

Communicative era	Key features
Oral culture	Orators Speeches Forums
Written culture	Scribes Manuscripts Libraries
Print culture	Authors Printed books Published collections
Electronic culture	Collaborative authors Digital documents and media Cloud and networked storage

of talk, text, and shared media. In the smartphone, one particular rem-
nant of telegraphic communication resurfaced—the short text message.
Similarly, in the electronic era, simple text messages and emails have
supplanted, to some extent, earlier forms of handwritten and printed
personal and business correspondence. Other examples of current technological forms, which include remnants from previous eras include the
use of microphones for presentation (oral era), stylus pens to write and
select items (written era), printers to produce paper documents as well
as physical three-dimensional objects (print culture), as well as digital
display screens (electronic) with haptic controls, allowing us to navigate
information environments in more complex ways.

The shift from one dominant communicative era to the next, such as
print to electronic, is often a longer, gradual transition over several decades or
centuries, rather than instantaneous. As a result, many hybrid technological
forms emerge during these communicative shifts. For example, Gutenberg's
invention of the printing press in the mid-15th century is often considered
to be the watershed moment for the print era and its resulting information
products, including printed books and papers. As a result, over time, new
technological innovations and notions emerged, such as mass production of
printed newspapers, copyright and trademarking, and subsequent improvements in public reading literacy. Subsequently, electronic forms of communication have been influenced by Babbage's invention of the programmable
computer or Bush's Memex in the early to mid-20th century, which has led
to newer communicative forms such as the personal computer, the Internet,
and the World Wide Web. The dominant shift from print to electronic
communication forms began in the latter half of the 20th century, and as
more electronic technologies and forms have become prevalent, hybrids
continue to emerge based on necessity, novelty, and use. In the 21st century,
information product environments have become increasingly more visual,
interactive, and semantic in their characteristics and forms. And while Ong
did not postulate specific communication eras beyond simplistic notions of
the electronic era, his work suggests other dominant communication eras
will eventually emerge beyond the electronic age.

Understanding Information Environments as Hypertext Systems

Hypertext theory underpins many of the feature and practices in creating,
developing, and presenting content in electronic information products.

These products include a vast range of applications, electronic documents, interactive media, software programs, and websites, which characterize the kinds of information products common to the electronic age. As previously mentioned, some of these key features include capabilities such as hyperlinking, flexible information structures (both linear and nonlinear), customization, interactivity, navigation, collaborative structured authoring, multimodality, and component content (Baehr & Lang, 2019). As information environments and their products have continued to evolve, these characteristics and features inform both development strategies and unique characteristics of information products today. To help better illustrate, each characteristic is described in the following sections in terms of their unique features and with examples of common practice and implementation.

The most basic and prominent characteristic or feature is the hyperlink, which allows content to be semantically linked by conceptual association. Hyperlinks allow users to customize access, order, presentation, relevance, and semantics when reading or using information products. They also offer users more complex possibilities for access and interaction, enabling them more flexibility in interacting, navigating, searching, and experiencing content. Hyperlinks can serve many purposes and functions, including the ability to outline content, show relationships, suggest concepts, or even indicate function (Baehr, 2007). For example, within a single web page, hyperlinks can serve any of these functions simultaneously. When presented as a list in a multilevel nested navigation menu, hyperlinks can outline the various information levels of content present. Hyperlinks can show relationships simply by linking two content items together, which suggests the presence of semantic similarity (Bolter, 2001; Landow, 2006). They can suggest concepts between linked content items by using descriptive keywords or images that directly relate to one another. And finally, hyperlinks can indicate function as clickable images or text, using icons, pictures, shapes, or textual descriptors that enable users to interact with the site's content. Whether they communicate structural patterns, relatedness, specific concepts, functionality, or use, hyperlinks facilitate a more active and unique exploration of an information product.

Hypertextual characteristics and structures accommodate flexible organizational patterns for information products and their environments. In structural terms, while linearity and nonlinearity are not necessarily binaries, together they describe the range of possible structural configurations for information products and their contents, whether linear, hierarchical, hypertextual, or customized in their overall structure (Baehr, 2007). Quite often, information products require a combination of multiple structural types, depending on the functionality, presentation, and volume of content. While linear characteristics were more the standard organizational pattern for most print-based and early electronic information products, they are more commonly used today for specific functions in information environments, such as guided or sequential processes used in instructional contexts or training. For example, in a website, the process of completing an order for purchases, or entering information to register for a course or event, might require a linear process helping users follow a step-by-step iteration. A hierarchical structure may also be somewhat linear, organizing topics from general to specific, but accommodate nonlinear exploration, allowing users to jump from one topic to another in hypertextual fashion. However, nonlinear components, such as reference hyperlinks or performance tips, might be included as supplemental content, which branches off from the main linear sequence. A combination of these organizational techniques may also be used together, adding layers of sophistication and complexity to product environments and information experiences.

Customization and personalization are also hypertextual aspects of information product environments. While product developers create the interface (or information environment), some interactive features may provide users with options to customize their experience. These customizations may include how content is accessed, navigated, organized, or presented. Customization options can be automatically provided by the system or enabled through user controls or settings within the interface. Within many content management systems, the user's interaction with content can also affect both the behavior and presentation of content. Whether content is customized by the system or the user, content options can be pushed (delivered automatically) or pulled (user enabled), which underscores both the importance of both user and interface in creating the overall information content experience (Lanham, 2010). In some cases, content (or control over the interface) may be both push and pull, and may be dependent upon features present. For example, in many commercial websites, lists of popular products may be featured on the home page, based on collective or individual user browsing history or purchases, or as featured or related sale items, which are pushed to the user when the home page loads. Users may also pull content by using a keyword search to access specific products relative to their interest or purchasing needs. While such a system may provide some customization options, users can also personalize their own information experience through both active and passive behaviors as content is pushed and pulled to them through the website's interface.

Interactive content is another hallmark of hypertextual environments, which accommodates more active engagement with an information environment. As information technologies have evolved in the electronic era, content has become increasingly interactive for users. Information environments can provide a blend of both static and interactive content, which can accommodate both stable content and dynamic properties, which accommodates greater immediacy, or a more customized and engaging information experience for users. Bolter & Grusin (2003) describe immediacy as a feature whereby the user can interact seamlessly with content in a way that the interface, or environment itself, is perceptually transparent to the user. This transparency is essential to creating immersive experiences, where users can comfortably perform critical tasks within an information environment. This seamless interaction allows users to explore information product environments, which may serve a wide range of purposes, such as demonstration, entertainment, instruction, presentation, and socialization. Within an information environment, not all content requires interactive properties; however, users have increasingly come to expect a certain level of interactive immediacy in their information experiences. As one example, games and simulations offer a wide range of interactive content forms, from short animated videos to multidimensional immersive environments. For example, web-based training can include interactive games or simulations, such as the ability to interact with virtual models, practice simulated tasks, take performance quizzes, and collaborate or socialize with other participants. These interactive properties are often facilitated by both system and user in a repeating chain of input (user interaction), processing (system data handling), and output (presented content) as an interactive chain (Crawford, 2003). As such, the perceived transparency of an information environment, or interface, may also complicate our perception of the system, whereby we may begin to see it as another user, of sorts, with which we can interact.

Flexible navigation is yet another key hypertextual characteristic found in many contemporary information product environments. While users may expect certain navigation functions or tools, such as nested menus or keyword search, these expectations may reflect their own individual experiences, motivations, and purposes. Flexible navigation should ideally offer users both options and variety, featuring a range of choices that can facilitate access, aptitude, preferences, and use (Baehr, 2007). Providing users with a range of navigation options can also accommodate their individual learning and performance modalities, which can be also considered a hallmark of good usability. For example, information wiki platforms often provide users with the ability to create, delete, edit, link, and notate content topics within the system, based on their own perceptions of content depth, semantics, and structure. As simple readers, users can explore content topics using tools such as bookmarks, hyperlinks, keyword searches, site maps, and others, which allows them to customize their reading experience, including the order, speed, duration, and level of interaction desired. Providing additional navigation options, such as a nested topical toolbar menu, a tag cloud of popular keywords, clickable directional arrows, scrollbars, and other interactive tools can provide users with addition choices to customize their information experiences. As content contributors, users may be able to create their own associative connections, through adding hyperlinks or annotated content, through commenting or editing, which accommodate flexibility within the system. Accordingly, users can choose their own paths or methods of interacting with an information environment that align with their own mental models and navigational preferences.

Hypertextual systems can also accommodate collaborative authoring, which can include both content and structural characteristics. Some information product environments may allow users to annotate, create, edit, structure, or even revise content and its presentation within the system. Creating content often involves both user and system, which function as counterparts, particularly in systems that evolve through such collaborations. Users may create their own content topics within an information wiki or in tandem with machine assistance, including use of algorithms, artificial intelligence, and other programmatic functions, which may automate or augment ways in which content is displayed, organized, and presented. Within an electronic information product, content development is much more than writing—it often includes both structural and semantic coding, requiring the creation of information architectures, component content units, navigation tools, and the use of markup, programming, and scripting languages that add both function and meaning as semantic layers to existing written content. While some content may be developed by human users, it can also be automatically created or generated by the system itself. Rockley and Cooper (2012) describe this function as adaptive content, which changes based on conditions generated by either the system or its human users. Adaptive content is a key feature of structured authoring, whether working in a content management system or hand-coded environment. The use of interactive scripts, such as algorithms, conditional statements, functions, or other system-generated means, allows content to be dynamic, which may change its properties based on conditions or user interactions with the system. These actions might include users entering textual content, selecting specific options, or even clicking on hyperlinks to trigger specific actions within the system. These conditional or reactive properties may be developed through hand-coding by a human user, edited from a library of existing coded scripts, or automatically compiled and generated by the system. Within a simple word processing program or web development platform, using stock templates or autogenerated code to create a new document could be considered a collaboration between user and system, working together to create new content. Similarly, this new content can be structured into chunks, sections, or new pages, but it can also be assigned semantic metadata, which adds another layer of depth to content, whether this information is encoded by the human user, the system, or collaboration of both. While some artificial intelligence tools can be used to create or generate such content, this collaboration between user and system may require additional development or revision in terms of its context sensitivity, to ensure the desired message is communicated.

Multimodality, as another hypertextual characteristic, suggests the hypermediacy (or multimediacy) of electronic information environments, which can include different (and multiple) content forms, such as the use of audio, graphics, text, video, and other interactive content (Bolter & Grusin, 2003). Multimodality may also include the use of multiple communication channels and modes within an information environment. Multimodality can often provide choices or multiple content layers as part of the information experience. As an example, training products, such as learning or educational applications, often incorporate multiple methods and modalities of presenting information, which accommodates learner choices in how content is organized or presented. These choices may allow learners to choose activities that focus on one modality, such as reading an article, listening to a podcast, watching a video, or performing a task, which accommodates their preferred learning styles in mastering a particular subject. In some cases, content may be presented in multiple modalities, such as an interactive demonstration, which incorporates audio, textual content, video, and task performance in a single presentation, which allows users to cognitively select modalities that best suit their individual needs. Providing multimodal choices or content forms can also support flexible methods of cognition and learning, or even added interest in information environments.

And finally, modularity is yet another hypertextual characteristic, which suggests that content within an information environment can be created and presented as flexible, reusable components (Bolter & Grusin, 2003; Landow, 2006; Baehr & Lang, 2019). Modularity suggests that content chunks within an information system function as reusable components, like building blocks, which can be repurposed, resized, and reused as needed. Within an information system, content can be stored as different component sizes, whether a single chunk or unit, such as a title, or a combination of chunks, such as an article, which might include a title, author name, timestamp, list of keywords, text, images, and reference links. For example, many e-commerce websites function as component content management systems, designed specifically to reuse and repurpose content from one or more databases. Content might be stored and retrieved based on keyword search, date range, product family, product number, cost, or other criteria, with the results repurposed and presented based on relevant entries. The notion of component content fundamentally changes the way we think about content authoring, as modular units, particularly in electronic information product environments where content may be reused for different purposes to create more customized information experiences for users.

In many ways, hypertextual characteristics continue to inform the development, implementation, and use of information products. Hypertext theory provides a fundamental foundation from which these products are authored, created, presented, and structured (Lang & Baehr, 2023). These characteristics provide a framework that can be used to develop collaborative content, information hierarchies, navigation tools, semantic relationships, and other features, which build upon hypertext's core theoretical concepts. While these concepts and characteristics provide foundational guidance, information developers, including technical communicators, must find creative ways in their adaptation and use within product interfaces, which ensure successful information experiences.

Environment Design and Development Challenges

Information product environments can offer a complex range of content forms and functions to support user interaction; however, sometimes their actual implementation can present unique challenges for developers. While hypertextual characteristics provide foundational guidance for developers, they must carefully consider how to customize these characteristics and features to meet both product specifications and user needs, and to avoid potential problems. Often, the information environment itself, rather than specific content assets, can be the root cause of specific development problems, which may be related to its accessibility, functionality, presentation, and perceived usefulness. The development of an information environment may be based on many factors, such as developer preferences, product specification, resource availability, technological dependency, and user expectation. As a result, information product development can involve many complexities when attempting to address these factors. Additionally, any shortcuts in the development process, such limited usability testing, a lack of developmental prototyping, or unchecked process automation, can potentially introduce errors and other problems that negatively impact the information experience for users.

Selecting the technological platform and the set of features that matches the unique needs of both product and user is one such challenge. While a particular platform may have a wide range of features and tools available, the developer must select only the features that best fit both product environment specifications and unique user needs. Resisting the urge to integrate features based on novelty, popularity, or trend can be tempting for some developers, but in the end, these choices may be perceived as superfluous, and negatively impact the information experience for users. Some platforms may offer automated features, standardized forms, templates, and even authoring assistance; however, sometimes these features can limit creativity and user-centered product development, without proper contextual inquiry. For example, novice web developers may rely more on standard design templates available in a particular platform or hosting service when designing websites. The use of these templates may help expedite the design process, but they are often limited in capabilities, customization options, and features. Web development platforms often integrate basic automated authoring tools and features to assist in creating basic feedback forms, navigation menus, search features, and other useful components; however, in many cases, they often require significant customization to appeal to specific user needs. In turn, unmodified features and templates may result in diminished information experiences for users. Experienced users may expect more complexity and customization in a typical information environment or product than is typically provided in standardized templates. Researching the full range of capabilities available within a particular platform, as well as both product and user specifications, can help developers make better choices for the ideal information experience.

Information environment development can present other complex challenges, mixing communicative features and modes, including content that can be both static and interactive, asynchronous and synchronous, and, sometimes, print and electronic. As such, these hybrid characteristics can create challenges in selecting the optimal navigation, organization, and usability features to integrate into a product interface. In some cases, development may require the inclusion of additional performance assistance or help features to assist users in both comprehension and use. In more complex information environments, users may seek out or even expect these assistive features. When these features are not present, users perceive the interface as unnecessarily challenging, frustrating, or restrictive of their needs. As information products evolve over future iterations and updates, developers should also consider how product interfaces and experiences are affected by changes in content, design, or organization, as any changes in the system or environment invariably affect changes in both content and use. However, once new features become familiar to users, these changes will be relied upon and expected by users.

Implications of Information Environments and Information Experience

Information environments can incorporate a wide range of content forms, hypertextual characteristics, interactive features, and other visual, spatial, and textual codes, which collectively function as an information product's interface. Interfaces can incorporate both semantic and structural characteristics, which communicate important organizational and relational concepts that help create engaging information experiences for users. By incorporating a wide range of features, interfaces can function as convergent hybrids, merging characteristics and features that include multiple forms of content, media, styles, and technologies into a singular product environment. Understanding how interfaces communicate these various aspects to users can help developers maximize the functionality and usefulness of information products and their related information experiences.

Considering some of these implications, including how interfaces support, and perhaps even enhance, content can help developers create more successful products and information experiences for users.

Interfaces connect users and content through a holistic, seamless experience. The interface facilitates the perceptual and cognitive link between user and content, which supports the overall information experience. These information environments present the content and framework of an information product, which includes any controls, design elements, interactive features, and tools present in the interface. Users may have some degree of control in the interface, depending on the customization options or choices present, including content size or styling, navigation tools, visual display options, and other aspects. These choices might include multiple navigation tools, hypertextual or customized structures, display or desktop customizations, advanced sorting and filtering options, and other configurations that support ease of use. Information environments can also feature both push and pull content—where some content is provided (or pushed) automatically by the system, while other content may require user action (or pull) to be displayed. Providing customization or control options for users can help them adapt and learn new information product interfaces more easily. When interfaces are perceived as easily used, or even intuitive in nature, they can provide a more seamless information experience.

Information environments integrate many useful hypertextual characteristics. While each information environment may be unique, many hybrid and electronic information products incorporate foundational hypertextual characteristics, features, and methods in their development. These characteristics include the use of hyperlinking, flexible information structures (both linear and nonlinear), content customization, interactivity, navigation, collaborative structured authoring, multimodality, and component content (Baehr & Lang, 2019). Since these features are quite commonly found in many electronic information product environments, users have come to expect many of them in their information experiences. Hypertextual characteristics support content with both simple and complex structures, incorporating linear, hierarchical, hypertextual, and even highly customized configurations. In turn, these information structures can enable and support more interactive and semantic connections within information products. The use of hyperlinks to create associative connections between individual content topics within an information structure can create additional possibilities for flexible browsing and searching of information. The use of hypermedia accommodates multimodal content, which can be combined, layered, or semantically linked to other useful resources within an information environment, which can include audio, interactive, static, video, or virtual content. These characteristics can support complex and dynamic content, which can be interactive, layered, personalized, or system-generated for different purposes and uses. Therefore, the range of hypertextual characteristics offer developers options for enhancing the interactive nature of content within a product's interface and to create a more engaging information experience for users.

Information environments function as convergent technological features and forms. While each communication technology may be associated with a specific era (oral, written, print, electronic), most are imbued with characteristics and features that are convergent or remediated from multiple technological eras. Convergent hybrids describe a set of technological products that overlap multiple communication eras and modalities. As one example, the electronic computer printer merges the capabilities of processing electronic documents into printed documents on paper, while the document scanner (when combined with a printer) allows printed documents to be converted into electronic formats. Similarly, as information products evolve, they often incorporate new features and forms based on both convergent and technological evolutionary trends. The pen, or stylus, has undergone many iterative development cycles, being used to write manuscripts in the written communication era, to edit or markup paper documents in the print era, or even to design or draw illustrations on the screen in the electronic era. And in some haptic environments, the pen is replaced by the finger, allowing users to navigate, select, trace, and interact with content presented. These convergent features and technologies have important impacts on information environments, specifically, as they affect the ways in which users interact with them. Technological convergence also affects how we develop information products to ensure they provide users with useful content interactions, navigation options, and information structures, which support a positive information experience.

Interfaces have languages of their own, which users must learn. Interfaces have their own language, of sorts, mostly in the form of visual, spatial, and textual codes, as well as conventions that communicate how content is accessed, customized, organized, presented, and ultimately used. Experienced users learn these conventions from one interface and apply their knowledge to learning and using others. In some cases, they learn through use, while they may require come contextual clues within the interface to aid comprehension and performance. Once users master an interface and its features, they can optimize their performance and use of an information product to suit their specific needs. Our combined structural and semantic understanding of any information environment is closely linked to our perceptual and cognitive processes, including how we learn new interfaces. Information environments that incorporate characteristics and messages that align with these processes can better support performance and use. They can incorporate specific semantic and structural messages through the interface, which supports the comprehension of content, function, organization, relationships, and use of features and functions. These messages may include the use of alternate descriptions, headings, help features, instructional labels, performance tips, tutorials, and many others. As such, these structural and semantic messages support important usability factors such as accessibility, findability, searchability, and usefulness. When users learn the language of the interface, this comprehension can lead to improved performance and use.

Information environments evolve both naturally and inevitably over time. As information technologies continue to change, converge, and evolve, information product environments will continue to follow suit. Technological products are not isolated inventions, and, in fact, successful ones inevitably converge into other iterations, forming new product features and experiences. For example, tablet computers, such as the iPad, combine affordances and features from several other useful technological products, such as the camera, desktop computer, interactive mapping software, stereo, telephone, television, and others. Information products, such as websites, began as mostly linear structures with static content pages connected by simple hyperlinks, yet evolved into more customized dynamic content management systems, with both user and system generated content. Each subsequent technological iteration incorporated both convergent and new features from previous forms, evolving into more complex information products. And since information environments represent the functional interface between user and content, this interaction will naturally evolve as well, through each new technological innovation and product iteration.

Chapter 5

Strategic Branding

Information experiences represent holistic messages, conveyed through various visual, spatial, and textual codes, which communicate these messages from information products to their intended users. Often these messages are carefully planned and executed by developing strategic brand and visual identity guidelines that help communicate these messages. Brands can represent specific themes, which are interpreted by developers through an information product, and communicated through specific characteristics, features, messages, and styles. Product branding establishes this messaging through both strategy and tactics, enabling developers to imbue products with specific supporting characteristics and features. An entire team of professionals may contribute to brand development, whether that brand represents a single information product, an entire product family, or an entire organization. Almost every role within the development team, including the project manager, content strategist, content provider, information designer, and others, contributes in some part to creating, implementation, and maintaining the overall brand.

In particular, information designers are often critical in brand development, creating the visual identity guidelines, style sheets, and content that best communicates the brand experience to users. Visual identities function as global style guidelines for an information product, which align with the overall brand strategy as well as established information design principles and practices. Specifically, these guidelines incorporate the practical aspects of design, including colors, icons, images, logos, styles, and other elements that help communicate the brand experience (Baehr, 2007). Visual identities represent the brand and are implemented through style sheets, which provide design specifications at every level of the information product's design. To be successful, a visual identity should also be informed by established information design principles and practices, including specific information design principles, design conventions, and style guides. In turn, a successful alignment between the overall brand strategy and visual identity can result in a more consistent and positive information experience for users. Consequently, branding and visual identity are closely related in helping developers and teams create consistent and effective messaging, which is integral to creating a cohesive information experience for information products.

Visual brands are also an extension of holistic visual-spatial thinking. Conceptual wholes represent the entirety of visual, spatial, and textual codes present in an information product. Our perceptual and cognitive processes help us interpret these codes individually, as well as collectively, as conceptual wholes, that communicate the branded messages of an information product. Our visual thinking helps us interpret information environments holistically, which includes our understanding of a product's overall branded messaging. Whether a particular brand represents cutting-edge technology, innovative experiences, modern design elements, or sustainable solutions, these characteristics and themes are communicated through the brand's unique messaging and visual identity, which support all other information product and environmental aspects, including its content and functionality.

Understanding Brands and Visual Identities

Brands serve as the strategic guidelines, while visual identities represent the tactical blueprints, or specific design and stylistic choices, that are used in developing an information product. While a brand focuses on larger messaging strategies, such as slogans, taglines, themes, and so forth, the visual identity focuses on the information design aspects using various visual, spatial, and textual codes. Successful visual identities are said to be on-brand, providing specific guidelines for designers, which align with brand messages and themes. While brands represent the overall content and thematic messages, visual identities incorporate appropriate information design techniques that support them. These supporting techniques should also be informed by established information design principles and tactics, which might include conceptual, consistent, contrastive, positional, and relational design techniques, which are described in a later chapter. Regardless of the specific techniques used, established design theories and principles can support designers in making more informed design choices and decisions.

Creating and implementing a well-branded visual identity ideally translates into a successful information experience for users. Brands and visual identities function together as a collective, conceptual whole of the information product experience in the user's mind. They also function as visual gestalts   communicating a holistic set of messages from information product to the user. These visual gestalts, as conceptual wholes, are formed by our perceptual and cognitive processes as we interpret the various codes and features of an information design. For example, if a design studio's brand messaging includes themes such as modern, open, organized, and welcoming, the visual identity guidelines should incorporate elements that complement these messages in both concrete and abstract ways. Some related applications might include the use of specific images of interior spaces with a modern design aesthetic, such as clean lines, minimalistic features, monochromatic colors, and open spaces. When implementing brands and visual identities, designs should also incorporate visual and spatial thinking in both design and development. While the visual identity guidelines and supporting brand strategy present one set of design specifications for the designer, these constraints are best filtered through visual-spatial thinking processes and principles, which govern how users interpret visual, spatial, and textual codes. Information designers can also think visually and spatially when creating and refining design concepts, using these established principles of design as heuristics in developing information products.

Successful brands and visual identities should also align in communicating specific messages and themes. Visual identity guidelines include the visual, spatial, and textual codes and conventions for product design and presentation, which communicate the explicit and implicit branded messages and themes that represent an information product experience. While visual identity guidelines focus on the visual and spatial content properties, they often enhance and can be enhanced by textual messages. For example, a visual brand that includes color schemes (complementary blues and oranges, sepia tone neutrals), images (mechanical icons, cogs, wheels), and typefaces (industrial or mechanical font faces) may support content that suggests a technological theme and its corresponding technical content and products. Supporting spatial techniques (horizontal and vertical rules, symmetrical grid layouts) can be used to position, organize, and group elements to create structural patterns that communicate well-organized spaces. And supporting textual content, such as keywords (mechanized, automated, intuitive) and themes (technological, industrial) can be communicated implicitly and explicitly through headings (system configurations, technical specifications) and slogans (trusted computing solutions). Together, these codes and messages support an overall information experience that communicates the information product brand to its intended users. And once an information product has been launched, it is essential to collect regular feedback from users to help refine, improve, and ensure continuing alignment between the brand, its visual identity, and the overall desired information experience.

Brand Discovery and Evolution

While branding strategies provide the guidelines, messaging, and themes for the blueprint of an information product, information designers are ultimately responsible for the development and implementation aspects of branding. The process from development to implementation often follows an iterative process, whereby specific design styles and specifications help designers create conceptually holistic, memorable, and successful brand identities. These design tactics are implemented through a set of visual identity guidelines, which specify the unique visual, spatial, and textual codes that communicate the brand's unique characteristics and themes. The process of developing a unique visual identity includes various phases and tasks such as brand discovery, research, design, and implementation (Merrilees, 2021).

Brand discovery involves the planning and development of specific brand messages and themes that best align with the target audience, or users, as well as the desired characteristics that best communicate the brand. Successful planning also encompasses researching the target audience, competing brands and products, or even relevant design practices, standards, or theories, such as how users think visually and spatially in how they interpret brand experiences and messages. Both brand discovery and research are important planning aspects that aid in the development of both characteristics and themes, which ultimately represent the product and its desired information experience. Brand discovery includes brainstorming the various goals, messages, outcomes, slogans, and themes, which conveys the intended information experience for product users. Often in brand development, product experience cards are used in brainstorming the overall concepts, keywords, and themes, which will represent the brand. These characteristics can vary widely, including keywords such as creative, flexible, organized, timeless, trendy, and many others. In some cases, a marketing team or existing brand strategy may be helpful in creating new branded messages and themes to be used in product development. Additional supporting research for brand development may also include discovery about the intended users and uses of a particular product. This research can be informed by what is known and discoverable regarding user behaviors, demographics, expectations, and habits to better understand how current brands are perceived and what expectations users have with other information products. This benchmarking research can help determine what characteristics, features, and themes are used in competing information products and experiences. Benchmarking research can also be useful in determining gaps in product development, which can help teams create a more competitive edge with a new brand.

Developing the specific design characteristics, features, and styles into a set of implementable visual identity guidelines follows the work of brand discovery and research. This involves the creation of specific design specifications, such as color palettes, grid layouts, and style sheets, which integrate consistent and creative techniques used to communicate the visual brand through various visual, spatial, and textual codes and conventions. In this particular phase, information designers are instrumental in developing the specific conventions, or visual identity guidelines, that function as a set of specifications that align design techniques with branded characteristics and themes (Baehr, 2007). Information designers must interpret the characteristics, goals, messages, and themes of the brand into the specific visual, spatial, and textual codes used in information product development. Visual identity guidelines function similarly to design style sheets, including the specifications on colors, images, interactivity, layout, media, spacing, and other style information that represent the brand. Design style sheets provide the conventions and specifications that information designers must implement into an information product, spanning both content and environment. These conventions can be applied locally, in one or more sections of a single page, as well as globally, throughout multiple pages or sections of an information product. While style sheets may include prescriptive rules for an overall design concept, the visual identity can also be more conceptual, including broader guidelines for interpreting the brand, such as offering choices for color palettes, imagery, layout, and other presentational aspects. A wide range of tools can assist designers in automating the development of style sheets, such as the use of scripting languages, development software, and content management systems. These tools provide developers with color scheme pickers, image libraries, templates, and other features that can be customized so they conform to a specific visual identity.

To further illustrate, Cascading Style Sheets (CSS) is the primary scripting language used in developing design style sheets in electronic information products, including websites. CSS allows specification for specific design features, including interactive, positional, and stylistic properties and values in presenting content. While other markup and scripting languages are useful in presenting and transforming content into static and dynamic forms, CSS primarily focuses on presentation of visual, spatial, and textual content. CSS is often used with other markup and scripting languages, which can provide additional interactive, semantic, and structural characteristics for information products. For example, when creating websites, Hypertext Markup Language (HTML) elements can be used to mark up content, which can include its semantic value, such as the use of headings to emphasize specific keywords or sections within a document. When combined with stylistic properties from CSS, the literal and semantic messages present within content markup can be enhanced through additional visual and spatial codes, such as color, shading, spacing, or other related properties. From a coding perspective, the combination of content and presentation markup and scripting demonstrates how visual, spatial, and textual codes work holistically to communicate successful information experiences for users.

Visual identity guidelines can include a wide range of design properties and values, which can specify the use of backgrounds, borders, color schemes, positioning, shading, spacing, and many other aspects of presenting content. They can also include specifications for various images used, which may include the use of logos, icons, drawings, photos, clipart, and others that span a wide range of graphic types, varying in both format and level of detail. Positional elements typically include the use of layering, spacing, and sizing to suggest varying levels of depth, emphasis, relatedness, or other semantic characteristics. Stylistic properties can also specify margins, borders, padding, and negative space, which can be rendered as static or responsive elements, based on specific requirements. Depending on the type of device, screen, settings, or other factors, responsive elements can be repositioned, resized, or even have visibility settings changed with the support of other markup and scripting languages. And within electronic information products, design elements can adapt or change based on user actions or specified conditions, adding an additional layer of complexity to the presentation of content. Within this wide range of presentational capabilities, information designers can implement features and styles that represent branded concepts and themes to evoke a particular feeling or tone within an information product. Design concepts are often built upon specific design aesthetics (or themes), which might be characterized as contemporary, modern, technological, traditional, or something specifically customized to the desired brand experience. In a sense, the visual identity guidelines represent this unique design aesthetic, which can be customized to communicate a particular brand and information product experience.

As an example, the Carbon Sitars brand is based, in part, on the instruments themselves, drawing inspiration from the clean, modern designs of the product (see fig. 5.1). The brand incorporates both modern and traditional thematic elements, such as the use of classical sitar images along with modern elements, such as the sans serif font, minimalist color palette, and clean lines used throughout the design. The individual instrument images illustrate the traditional features of a sitar, such as its base shape, stings, and tuning knobs, as well as modern elements, including the use of minimalist backgrounds, feature icons, high-contrast techniques, and instrument materials and textures featured. The juxtaposition of the floral background against a high-gloss surface on one of the instruments featured also supports the combination of modern and traditional elements in the design. The brand logo incorporates consistent and well-aligned design features, such as the atomic rings around the letter C, which supports both the product and element name, carbon, which incidentally is also used in the product's construction. An encapsulated version of the brand logo simply uses this singular design element, which supports the brand identity. This unique design feature also supports the modern theme, aligning with other elements present in the design.

Collectively, the design elements used to communicate the branded themes demonstrate consistency, such as the use of both colors and logos, as well as creativity, through the combination of classical sitar features and minimalist design techniques. The product design composition demonstrates versatility, whereby different components of the design could be used to successfully communicate the brand and its products in a wide variety of information products, including brochures, marketing copy, product

Figure 5.1. The Carbon Sitars brand identity. The Carbon Sitars brand incorporates both consistent and creative elements inspired by the instruments themselves, including the unique design features and materials used in their construction. *Source*: Carbon Sitars, Brand media kit, 2024. Used with permission.

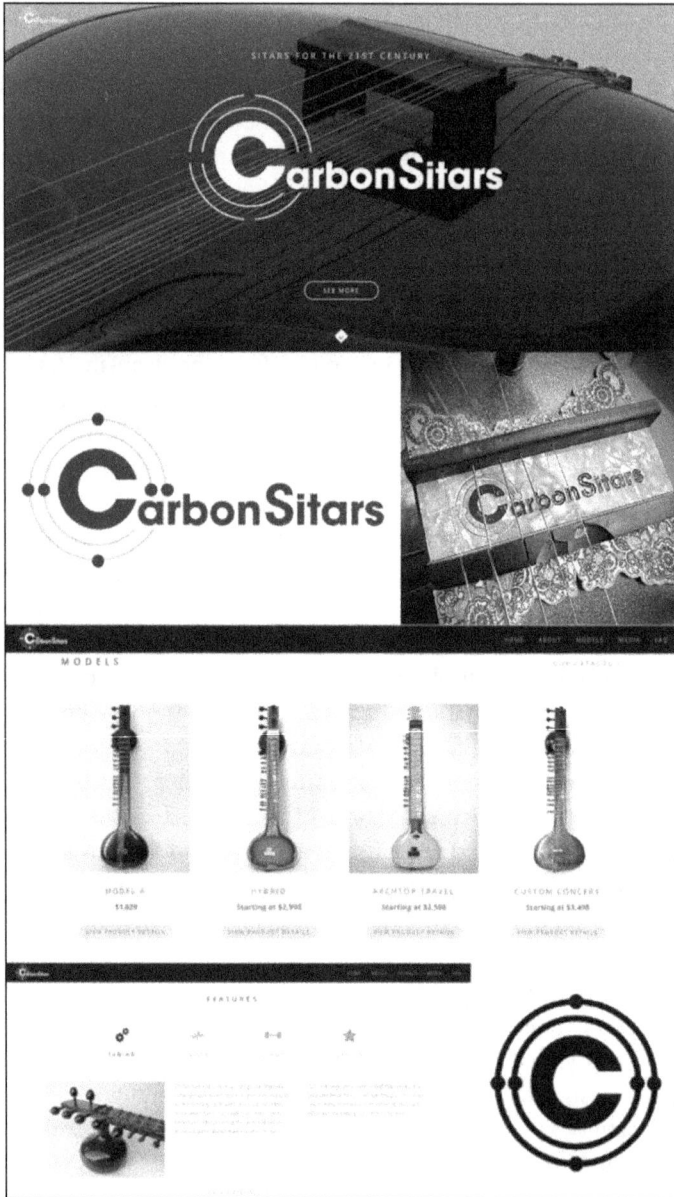

packaging, and websites. Throughout the range of these products and the unique design features used, the brand can communicate the modern and traditional themes as part of the information experience for its users.

Establishing compelling, recognizable, and versatile branded designs often requires a combination of both consistency and creativity, and the use of visual identity guidelines helps establish these characteristics in a holistic brand identity (Airey, 2019). Consistency ensures similar sets of visual and spatial styles and techniques are used, such as color schemes, icons, images or photographic treatments, symbols, and so forth, which can support brand recognition among product users. Consistency, as a well-researched principle of design, supports the creation and use of specific design conventions and standards, which can be implemented throughout an entire information product design. Within consistent design standards, there must also be flexibility in the visual identity guidelines to allow for creative approaches to the implementation of both brand and design. While consistent design specifications establish stable and successful visual brands, they must also be adaptive and flexible to also accommodate brand evolution and freshness. Creativity expands the application of visual identity guidelines beyond a singular, rigid application of both positioning and styling. Working within a set of constraints can also sometimes foster creativity in design work, making it easier than working with a blank slate, or completely standards-free environment. For example, a set of visual identity guidelines might include the use of warm colors (reds, oranges, yellows), sunshine or sun-shaped symbols and icons, sans serif typefaces, and balanced rectangular grids. While there are specific guidelines for color, layout, symbology, and typefaces, there is enough variability for creative application. Within these limitations, designers can use these as a baseline from which they can build upon the brand, developing other design elements that are complementary.

Visual identities, and the brands they represent, should be consistently dynamic enough to create a memorable experience for their intended users (Airey, 2019). For example, consider how a digital artist is commissioned to create a brand logo for an information product, using a limited set of colors (greens and blues) and shapes (squares and triangles) as design guidelines. While this is initially a seemingly broad set of guidelines, the designer works within constraints given to develop creative prototypes of the logo to present. Additional specifications may be provided after reviewing several prototypes, such as restricting colors to specific hues or shades, which often helps guide the creative process, by setting additional limitations for the designer to work within. Imposing too many constraints may stifle the creative process or innovative design approaches. Therefore, an iterative design process might include specific constraints and standards while allowing for some creativity in their application. To achieve such a balance, the development and implementation of visual identity guidelines depends on both principles of consistency and creativity to communicate a product's unique brand.

Brand implementation involves transforming information design concepts into working prototypes and fully realized information product designs. The implementation phase involves the actual work of creating prototypes and design composition layouts, using the visual identity guidelines in creating the specific design elements and products, such as banners, images, logos, pages, styles, and so on, which complement all forms of information product content. Developing design concepts and prototypes for evaluation and review may also include various forms of testing, such as gathering feedback from expert groups or product users. From the various graphics, images, layouts, styles, and other elements used, brands communicate messages and themes through these unique design aspects, which may include specific information graphics, design layouts, product packaging, and product deliverables in their various forms. Information designers must also rely on established design principles and practices to inform the implementation of visual identity guidelines. As part of the implementation process, designers may test various prototypes to evaluate the information product design in a variety of conditions and settings. This phase of the design process is also iterative, where a particular visual brand evolves incrementally with regular feedback loops from the team and its users, through multiple design iterations. Rarely will any design concept be fully realized the first time through, as both the product and the visual identity guidelines are revised for optimal use to ensure they successfully communicate brand messaging and themes. Testing may provide designers with unique discoveries that cause them to rethink specific choices or directives created in previous phases of their work. Accordingly, the process may involve several repeated cycles of designing, prototyping, and testing. For example, a chance discovery during the implementation process, such as experimenting with use of space or color, may lead to making changes in the overall visual identity guidelines established during the design phase. In turn, this may also lead to adding new branded themes or design conventions to optimize the product's design.

Visual identities and brands are continually refined, upgraded, and improved through innovative ideas over an information product's life cycle. Brands and visual identities evolve and may need to be refreshed over time, based on any number of factors. Any changes in the product brand should be thoroughly researched, developed, and tested with users to ensure they meet adequate expectations. Once a brand and its products are well established, it's essential to build future iterations in brand upon what users expect (what features they consistently rely on and prefer) as well as what they need (new creative features that support them in some way). A lack of innovation and improvement may result in an unfavorable impression, such as a brand that appears dated or stale over time, which may in turn cause users to seek out newer and more useful products. Conversely, frequent and drastic changes to information product designs that fail to incorporate relevant user research, can potentially alienate users and cause frustration due to seemingly unnecessary changes in product design. While a design team may be eager to try out new concepts, without proper research, these changes can be detrimental to information product brands. Once customers have been lost, it can be difficult to reestablish brand confidence.

How Brands Support Information Experiences

When you think of a specific information product, such as a popular website, application, or streaming service, you may be acutely aware of some the aspects of its overall brand. That brand may be conveyed through its color schemes, images, logos, slogans, and textual messages. Brands may also be communicated through its perceived characteristics such as intuitiveness, organization, or usefulness. Successful brands establish a coherent messaging strategy, through both content and presentation, in ways that are both explicitly and implicitly communicated to its intended users. In fact, information product brands rely on the supporting visual codes and messages just as much as its textual ones (Airey, 2019). Ideally, product messages are carefully planned and targeted to convey specific characteristics, features, and themes, which can be easily recognized and remembered by users. Branding strategies often involve conducting a product experience analysis, where specific branded themes are developed and eventually integrated throughout an entire information product. These branded themes might include what the current product represents from both customer and developer perspectives, as well as forecasting future aspects of the intended information experience. Ideally, branded themes help guide developers in creating implementable characteristics, features, functions, and other tactics that support the overall intended branded messaging. These tactics also directly inform the visual identity guidelines of the brand, down to the information design and stylistic levels to include the use of color schemes, images, logos, and other visual elements. Therefore, a successful branding strategy must align themes with visual identity guidelines, which helps create consistent messaging throughout an information product experience.

But what exactly makes a good or successful product brand? Many successful name brands can convey both positive and negative connotations based on consumer experience, but they often begin with and are supported by targeted brand messaging and supporting visual identity guidelines. Initially, the brand is the first impression between a particular information product and its user, but over time, it can represent a more sustained reputation, from a single product throughout an entire product line. Brands must also communicate themes consistently throughout its visual and spatial elements that support its textual content. For example, product icons, logos, and slogans often integrate consistent design techniques with subtle variations that communicate branded messages in similar ways with creative variations (see fig. 5.2). Key elements used in the Carbon Sitars brand, such as the use of the product name, logo,

Figure 5.2. The Carbon Sitars brand logo. The Carbon Sitars brand elements include various uses of its logo, supported by consistent design elements and features, which align with the messages and themes used throughout their entire product line. *Source*: Carbon Sitars, Brand media kit, 2024. Used with permission.

slogan, and variations in design presentation demonstrate both creativity and consistency throughout their various uses. The product name uses a consistently repeated sans serif font, in various sizes and colors that reinforces brand recognition. Regardless of its spatial or visual styling, the name is used consistently throughout the various design compositions. The logo is comprised of two consistent elements, an atomic structure icon using the letter C and the product name. Subtle variations in color, spacing, and presentation are used throughout the brand, and in some cases, the iconic C image is used in place of the product name. The slogan "SITARS FOR THE 21ST CENTURY" communicates the blend of traditional and modern themes, combining an 18th-century instrument, the sitar, with the 21st-century message tag. Additionally, while images of the sitar convey primarily traditional messages (the instrument's shape and features), the supporting design styles (colors, fonts, and icons) focus more on modern elements in the design. The use of design elements in the brand also communicates the same blended message featured as the product's slogan. And to add creativity to the brand, positional and stylistic variations provide heterogeneity and added interest to the design while maintaining brand cohesiveness throughout each iteration and use. Collectively, these techniques communicate coherent messages and themes that support both the intended modern and traditional elements of the brand.

Information products can also convey multiple (and mixed) messages, such as affordable, dependable, high quality, intuitive, modest, and satisfying. Collectively, these themes are communicated through various design features, representing the overall brand as well as its intended information experience. While users form their own unique expectations, interpretations, and preferences through their interactions with information products and designs, the branded messages and themes influence this interpretive experience very deliberately. For example, successful branded experiences may include features that are sleek but not too flashy, cleanly designed, intuitively useful, easily navigable, or even simple, yet well-organized, depending on the product itself. Users may even prefer or seek out elements that fit their expectations when choosing and evaluating products relevant to their needs. Product designers should also consider how to address these unique expectations and preferences when developing designs elements, which are both on-brand and user-centered. As such, designers must successfully translate both branded messages and user expectations (or perceived interests) into implementable design practices and product features. Designers must be both creative and consistent in their development of branded features and messages in order to create more useful and highly desirable information product experiences for users.

When implementing branded messages and themes, it is important to remember that brand messaging is communicated to users through both its content and presentation. While a significant part of a brand message may depend on explicit textual messages, the visual and spatial components are equally important, both supporting and enhancing those messages. For example, consider the following brand slogan for an information product's technical support forum: to provide cutting-edge crowd-sourced technical support for the intuitive and modern user. Some examples of themes that support this slogan might include contemporary, cutting-edge, intuitive, modern, novel, technological, or useful. Some of these themes may be explicit, while other more implicit to the brand. While the branding focuses on both the explicit and implicit messages and themes, the visual identity supports the specific information design components and techniques that support them. To illustrate, the technological theme might be depicted using images of mobile computers, illuminated displays, snippets of computer code, or technological shapes and symbols. The modern theme might be conveyed using more minimalist color palettes, sans serif fonts, and design layouts that incorporate the use of well-positioned regions of negative space. These specific visual identity guidelines interpret themes down to the stylistic and positional levels of information design to support the overall information product brand. Ultimately, these choices may rely on a wide range of factors, such as the designer's specific interpretation of themes, design principles, style guidelines, and user research.

Brands can build their effectiveness by integrating research and techniques supported by visual and spatial thinking in their development. Visual identities may rely heavily on visual codes, but also integrate spatial and textual characteristics that help establish the overall design of an information product. Visual identity guidelines incorporate both stylistic (such as color, font, or shading) and positional (such as horizontal, vertical, or grid-based layouts) characteristics that function as a complete set, establishing a holistic (gestalt) sense of design for an information product. Often, an electronic style sheet implements the visual identity through its use of declarations, properties, and values for each stylistic and positional feature of an information product. These may include different categories or styles of design elements used, such as colors, heading levels, images, spatial grids, and typefaces. Both individually and collectively, these visual elements when used consistently and repeatedly in similar patterns or configurations, become recognizable elements that users associate with the brand. When executed successfully, the visual and spatial elements that comprise the visual identity will complement and enhance brand recognition and messaging.

Information design practices also incorporate a wide range of design concepts, conventions, guidelines, and style guides, which can help designers convey and develop a successful visual brand. Along with these resources, information designers must also consider the unique contextual factors for an information product, as well as the knowledge of how users think visually and spatially in interpreting brands. Brands may also precede and evolve concurrently with their associated visual identities, as various design styles, images, and desired themes change over an information product's life cycle and development. Successful branding also helps establish the relationships and interactions between a company, its information products, and users. Specific branded messages may also influence how users relate to and remain loyal to a particular brand over time, including future interactions with a product. Therefore, brands also function as implementable and powerful communication strategies, which support information product experiences.

How Visual and Spatial Thinking Supports Branded Messaging

The visual identity of a brand functions as the holistic combination of visual, spatial, and textual codes, which communicate specific messages and holistic information experience to users. While the visual identity focuses primarily on the visual and spatial aspects of the brand, these styles support the textual codes and messages in many important ways. For example, a brand's specific textual messages may include keywords, slogans, or taglines, which are incorporated into banner and logo designs, further enhanced by visual and spatial elements such as backgrounds, colors, images, shading, spacing and typefaces. While the textual message may be succinct and memorable, the visual and spatial codes support retention and recognition of those messages with users. Conversely, poorly designed messaging can easily convey negative impressions of the brand, causing its products to be dismissed or ignored by users. Both visual and spatial codes are often as important textual ones, particularly when user perception and cognition are essential components in interpreting brand messaging and overall information experience.

Visual elements convey their own messages, independent of textual ones, whether explicitly through information graphics or more subtly through thematic imagery. Visual icons, images, logos, and shapes are more quickly noticed, recognizable, and remembered for their concepts as perceptual shortcuts to stimulate recognition (Arnheim, 1997). More subtle visual elements, such as the use of negative space, shading, or watermark imagery may support a specific pattern, theme, or tone, which support more concrete concepts conveyed. The related visual thinking principle of *vision is selective* describes how our perceptual processes help us determine both priority and importance of what we see (Arnheim, 1997). As such, visual elements that are distinctive or relevant to our specific information needs are often noticed before others in the same visual field (Baehr, 2007). In any visual environment, our area of focus follows a visual hierarchy, where our perceptual processes classify visual elements in terms of their distinctiveness, position, usefulness, and priority that may be relative to other visuals (Baehr, 2007). In other words, we notice visual information that is well organized, placed, presented, and useful before that which is less distinctive or useful. Our visual processes may also be hindered in information environments that are less organized, overpopulated with visual content, or poorly presented. As a result, we may perceive their information experiences as less engaging or useful. For example, feature creep is a common problem in mature information products whereby each new version continually adds new features, as a means of upgrading or improving the product, regardless of their actual usefulness (Norman, 2013). However, users that prefer a more streamlined and optimized interface may become frustrated over time as simplicity is replaced with unnecessary complexity in information products. For example, a website that incorporates a grid layout to organize announcements and events may begin to look like a patchwork quilt as several new items are added to the layout, resulting in a cluttered and confusing design layout. Furthermore, our senses can become overwhelmed with excessive contrast or emphasis issues in the overall visual layout. When the same strong contrast design technique is used repeatedly, we may have problems classifying and sorting important or useful elements from others present. Therefore, information designers must consider how visual elements support both brand design and ways in which users perceive, process, and prioritize visual information naturally.

Spatial elements also communicate their own unique messages while supporting both visual and textual ones, as well as the overall brand.

Spatial codes may communicate concepts such as closure, continuation, disparateness, emphasis, relatedness, and similarity, which are recognized by our perceptual processes (Koffka, 1935). These concepts can be communicated through the use of grids, patterns, repetition, spacing, and other techniques, which frame, group, and separate elements in an information environment. These spatial characteristics communicate how information is grouped, organized, positioned, and even sequenced, in the same space or along a timeline. For example, in an information graphic that compares different models of laptop computers, spatial techniques might be used to evenly position elements, such as images, categories, and individual specifications, which suggest equivalent comparisons for different characteristics, features, and ratings. Spatial codes can also communicate different layers of messaging, which can enhance the presentation of visual and textual information present. As such, spatial elements are fully capable of communicating conceptual, multidimensional, semantic, and even temporal relationships present in information environments.

Visual and spatial codes often function as symbiotic elements in information design and particularly well in electronic and web-based information product environments. Visual identity brands can be more economical and easier to develop, implement, and maintain these environments as well. The availability of software and application-based tools often help automate and develop many aspects of digital information design. As an example, often these products include design templates and image libraries, which can be easily modified. While these affordances may save time in the design process, when used out of expediency and not adapted properly to visual identity guidelines, they can cost both quality issues and time lost in support and redevelopment costs. For example, templates often bring together the visual, spatial, and textual codes into a single, unified design. However, design templates can be both helpful and problematic to information designers. These stock templates are often based on generic themes or design concepts, which may be useful to novice designers but easily recognized by experienced designers and users. A lack of sufficient customization and modification of a template may communicate a lack of creativity or mediocrity, which may be misaligned with both the brand identity and overall information experience intended. While these templates may simplify information design work, they should be used to extend the brand's creativity through appropriate modification, rather than substitute for it. As a result, users may assess the value of an information experience and attribute positive characteristics based, in part, on its overall level of creativity and originality in design.

We also rely on our visual and spatial thinking to master both subject matter content and the environment in which information is presented. For example, think of how an interactive video-based tutorial integrates a combination of visual, spatial, and textual content to establish brand and also instruct users. Visual content includes the video animations of speakers, slides, animations, information graphics, images, logos, color schemes, and other visual information that supports both brand and instruction. Spatial techniques may include how the video presentation is organized with overlapping images, content windows, screen layouts, slide transitions, and positioning techniques, which support both instructional usability as well as branded themes. Spatial characteristics might also include how users can navigate along the timeline of a video presentation, incorporating navigation tools to control the pace and presentation. Textual content includes both audio and text within the instructional tutorial, whether about the tutorial's subject, the presentation environment, or the overall product brand. These different codes may complement each other in the same information environment (or space) through different combinations, methods, and modalities. As a result, these complex combinations can be configured to appeal to a broad range of learners, learning styles, and user preferences. And within any instructional product, users learn both content and environment in order to have a holistic, successful, and well-branded information experience.

Complementary visual, spatial, and textual content can support our comprehension, relatability, and use in interpreting information products and their unique brands. Holistically, we perceive these elements independently and interdependently, interpreting both individual and collective units to form meaning (Kohler, 1947). For example, when using a set of consistent and familiar icons to represent various concepts or functions in a design layout, such as an hourglass, house, a question mark, and directional arrows, each supports our understanding of their various uses. While some icons may have a singular meaning, others may have multiple meanings. Experienced users of websites may easily recognize an hourglass placed next to a text box as a search function, based on prior learned experience. However, a page (or document) icon may suggest multiple meanings, depending on the context of use. Often, this symbol may represent a drop-down menu on mobile versions of a website, but when used on a page with multiple document downloads, it may represent individual documents. Some symbols may appear to be more abstract in their meaning. For example, directional arrow icons could function as fast-forward and rewind options in media controls, or, in a website, function as simple previous page and next page navigational buttons, allowing users to move back and forth between pages or sections.

When these symbols are paired with text descriptors, such as the words *search, home, menu, help, forward*, and *back*, the complementary nature of image and text may aid user comprehension of both meaning and function. Similarly, when these icons are paired with section headers, they might function as advance organizers that can help readers comprehend not only the structure of a document, but the location of its contents, as well. Our visual and spatial thinking interprets icons and symbols as specific concepts that support our understanding of elements and environments (Arnheim, 1997). Therefore, aligning the use of visual, spatial, and textual elements can help ensure that specific messages are correctly interpreted by our visual and spatial thinking processes.

In some cases, the environment can support brands and messages through the use of various visual, spatial, and textual codes present. McLuhan (2017) suggests the medium, or environment, itself can also be part of the message, particularly in electronic forms of media content. The medium, or even media type or format itself, has unique features that can augment, enhance, and perhaps influence the overall perceived meaning of messages. For example, information product environments may have different visual modalities, which can alter the presentation and experience for users independent of content, whether that environment is an animated demonstration, interactive application, video presentation, or other form. In a website, scrolling message banners often used in headers feature interactive, enhanced messages. These banners can incorporate multiple messages, presented in a sequence, which incorporate changing backgrounds, hyperlinks, images, and messages in a single space for users. Similar forms of interactive content may illustrate a process in real time, displaying a step every few sections, to support synchronous task performance for users. These interactive demonstrations may even include user controls to fast-forward or rewind portions to further aid performance. When the interactive elements are replaced instead by a static page displaying the images vertically, the message changes when it loses the characteristics that each corresponding medium affords the design. Our visual and spatial thinking interprets the function and usefulness of the static versus the interactive versions quite differently as well.

While interactive environments can create a fundamentally different experience for users, they may not always be the most useful approach for every information product or component. While they can create a more engaging experience, this can also present a unique set of challenges for information designers with limited resources or skills. However, environment design underscores the importance of the medium itself in communicating brands, content, and information experiences for users. With these new experiences come new ways of thinking both visually and spatially. Through various perceptual and cognitive processes, users construct holistic representations of an information experience, including their unique impressions and perceived uses. Accordingly, this visual and spatial thinking supports comprehension of information products and their brands as functional, holistic design concepts. With deep foundations in Gestalt theory, visual and spatial thinking encompasses the acts and processes by which humans form conceptual wholes from the various characteristics, codes, and messages present (Arnheim, 1997).

Ideally, strategic brands, and the visual identity guidelines that support them, are informed by established design theories and practices. Interpreting these brands involves both perceptual and cognitive acts and behaviors, which focus specifically on what users notice, how they solve problems, how they discern semantics, and how they form concepts (Arnheim, 1997). Some specific information design concepts that are critical to our interpretations include the use of figure/ground contrast, visual context, conceptual shapes, and continuation and closure.

Figure/ground contrast helps users interpret the semantic relationships between objects (figures) and the environment (ground) in which they are presented (Koffka, 1935; Kohler, 1947). Figures might include individual images, photos, boxes, or other objects placed in the foreground of the design. Grounds represent elements in the background environment, which may include borders, images, patterns, shading, spacing, or watermarks upon which figures are placed in the design. Both figure and ground elements operate as different layers within an information design, which can be static or changing. Layered elements exist in some cases in both figure and ground—multiple figures and grounds can occupy the same or different layers within a single design layout. Depending on their unique design characteristics, figures and grounds also function as single units that convey their own semantic messages, such as relatedness or disparateness, prominence or subtlety, or even blended or balanced presentations. Figures and grounds can also function holistically to communicate the overall brand experience to users. Therefore, figure/ground contrast can be a useful design technique, helping users comprehend relationships between objects and layers within an information environment.

Visual context is another aspect of information design, which helps users discern importance, meaning, and organizational patterns within an information environment. Visual context is often provided through specific visual cues, such as the use of contrasting visual styles, spatial configurations, or pairing of text with visual codes for emphasis or enhanced meaning. Our visual perception helps us interpret these visual cues and then classify and organize these objects into a visual hierarchy that helps us interpret new information and prioritize what is important (Baehr, 2007). For example, humans tend to notice high-contrast elements before all others in a visual environment, as well as how those elements relate to others in that environment. Other elements present in the environment are classified and prioritized based on their distinctiveness, importance, position, semantic value, or other visual codes (Arnheim, 1997). While we may notice familiar or visually distinctive elements before others in an information design, eventually, our perceptual focus may shift back and forth between the various elements, depending on their relative importance and relevance to our specific needs or motivations at any particular time. The visual context in which elements are presented can help emphasize specific elements within an information environment and also assist users in classifying and prioritizing what is emphasized, noticed, and discounted.

Shapes and symbols often represent specific concepts, which can support understanding of a visual brand. Whether these shapes or symbols are universally recognized or brand-specific, they can be powerful design elements that communicate the semantic messaging of a particular brand. These might include the use of logos or icons as recognizable concepts, such as an hourglass icon for a search function, or even grouping elements using spacing or shading techniques to suggest relatedness between different elements. For example, in a travel website, the use of familiar icons or shapes in page headers or navigation tools, such as airplanes, cars, notice boards, suitcases, tickets, and so forth, might suggest various modes of transportation, travel schedules, or rules and regulations. The same site may also incorporate other shapes that represent various functions, such as a human icon for customer service, an information symbol for frequently asked questions, or an hourglass symbol for a keyword search. Other symbols might include variations of company logos used repeatedly for branding and recognition. While each symbol may suggest a different concept or function, together they represent the core functions of the business as well as the company's overall brand.

Continuation and closure are two other Gestalt-based principles, which can help communicate brand messages. Continuation and closure are based on the notion that perception helps us see partial views of an object as a complete whole through extrapolation, rather than an abstraction (Koffka, 1935; Kohler, 1947). To illustrate, the principle of continuation suggests that when we see part of a pattern, we perceive its continuation beyond the borders or limits of our visual perception (Koffka, 1935). Consider the use of a repeated image as a background or pattern on a page, such as a white picket fence used as a decorative background for a real estate company's website. While we may only see part of the fence, stretching from the left to the right margin of the page, our perceptual processes perceive it as a continual image beyond the borders of the page. Similarly, the principle of closure suggests that when we see part of an object, which may be partially obscured in a visual environment, we extrapolate it as a conceptual and comprehensible whole (Koffka, 1935). Both continuation and closure suggest our perception fills in these gaps or limitations, allowing us to form concepts or meanings from what information is visible, and previously encountered. An image of the moon partially obscured by clouds in the skyline conveys the concept of a full (or whole) moon as a recognizable object, as long as sufficient contextual details are present. However, if there are insufficient contextual details or excessive abstractions in the image, we may be unable to perceive the correct (or intended) conceptual meaning. These conceptual extrapolations are only as successful as the surrounding visual and spatial cues provided for our understanding. Continuation and closure can help designers build techniques that help users interpret abstractions within an information environment, or even allow them to layer multiple objects in the same space, while still communicating branded experiences successfully.

Collectively, information environments present users with unique configurations of visual, spatial, and textual codes, which they must interpret and interact with in order to understand both brands and information experiences of products. Often, solving problems and completing tasks are essential to information product experiences, whether it's learning to use navigation menus, searching tools, filling out forms, interpreting an information structure, or sorting content results. Our interactions with information product environments depend on our perceptual and cognitive processes to help us successfully complete these tasks (Arnheim, 1997).

Visual branding and visual-spatial thinking work together to help information designers implement successful techniques that align user perception
and branded messaging strategies. These two elements also share a similar
function—to help users form a holistic understanding of an information
product's unique brand and its overall information experience.

Strategic Branding and Information Experience Pitfalls

Sometimes, brands and visual identities are less successful, perhaps
communicating unintended messages and experiences that differ from
planned intentions. Depending on circumstances, a poor information
product experience may be the result of poor planning, development,
or implementation of brands, design, or even messaging. While brand
development is largely strategic, its success also hinges on the tactics
used to communicate intended messages and themes, which are often
scoped out by the visual identity guidelines and specific product content.
Sometimes, strategy and tactics can be misaligned or misinterpreted by
both designer and user, which can create unexpected results in both the
use and perceived information experience.

As part of the planning process, strategic branding exercises often
involve the development of specific characteristics, features, messages, and
themes. While some guidance may be explicit and specific, such as using
a particular company logo and color palette, others can be more open to
interpretation, such as creating a product with an intuitive and technological appeal. To further illustrate the variability in strategic branding,
consider a website that features web hosting services, in which the brand
includes themes such as accurate, comprehensive, intuitive, and technological. In addition to these branded themes, specific branded content and
provided details include a product logo, an image library of screenshots, a
tagline and slogan, and a complementary color palette including specific
blue and orange color values to be incorporated. While some of these
elements provide more specific guidance than others, product developers
may interpret and implement some of these guidelines in ways that were
not necessarily intended. For example, the information designers could
choose to incorporate other images to represent the technological theme,
which seem unrelated or off-brand to users. Additionally, other development methods, such as the use of user research and usability testing
results throughout product development may create additional problems

or be misaligned with branded messaging. Periodic reviews and testing throughout development may help ensure that strategic brands and messages align with implemented tactics. Design team reviews can be conducted to ensure brands and messages follow appropriate guidance throughout each iteration of product development. Also, external user testing throughout iterations of product development may help determine how successful brands and messages are communicated from product to user, and help identify what adjustments must be made so the information experience aligns successfully with its intended brand.

Product brands may also be communicated poorly in other tactical applications, such as when the visual identity guidelines fail to support established and implementable practices of design, which may include how users think visually and spatially. In some cases, developers may ignore these guidelines, implementing styles that conflict with branded themes or even established design conventions and principles. For example, a key feature of product brands is the use of primary and secondary sets of branded colors. These color schemes may be used throughout multiple deliverables, such as websites, reports, instructional materials, and other technical documents used to feature a particular information product and organization. Using colors outside of the branded palette, such as different background or shading techniques, may convey incohesiveness or a lack of professionalism. In other cases, on-brand elements that fail to incorporate established design principles and practices may convey similar negative messages. Consequently, when developers create visual identities, style sheets, templates, and other visual content for products, the implementation of branded elements should both include a balance of brand specifications and adhere to established design principles and practices.

Over time, brands may also become stagnant, which can suggest other unintended messages such as staleness or other perceived lackluster characteristics. While some characteristics and themes may benefit from consistency, brands may require some evolution throughout the information product life cycle, starting with product development throughout future product iterations. When brands fail to demonstrate both creative evolution and consistent features, they may result in unfavorable information experiences for users. Users may come to expect certain features; however, developers can also introduce upgraded features and elements that improve and innovate brand messages while keeping the brand fresh and interesting for users. However, too much change in a brand can create information experience problems, particularly when developers frequently update designs that are motivated by novelty or the latest trends. Rapid changes or inconsistencies in information experiences can be perceived as confusing, frustrating, unusable, or worse by users. When products are redesigned, there should also be some degree of consistency from the previous iteration to ensure brand recognition and maintain product loyalty. Over time, users depend on some degree of consistency in information product experiences, particularly with features that help them browse, navigate, search, or perform other critical functions. Accordingly, developers should strive for a balance of both creativity and consistency when refreshing brands and designs, including any relevant user feedback (Airey, 2019).

Information products include a wide range of content, features, and uses, which all communicate messages that are critical to the overall perceived information experience. While a product's technical content is the substance of any information product, the branded elements incorporated into the product act as supporting thematic messaging. While technical content may be useful and reliable, a poorly conceived and executed brand strategy can compromise the usefulness and value of any product. Strategic branded messages have important communicative value, which ideally enhances and supports an information product to create favorable information experiences for its intended uses and users.

Implications of Strategic Branding and Information Experiences

Strategic branding provides information product developers with useful characteristics, messages, and themes, which help them align both content and presentation in creating successful information experiences for users. From strategic branding information, developers can create visual identity guidelines, which help with the implementation of various visual, spatial, and textual codes and content throughout an information product and its environment. When strategic messaging and visual identity guidelines are properly aligned, information products can communicate a well-branded experience for a product's intended users. Ideally, these two related practices are informed by established design principles and properly researched user expectations. Whether information products are physical, hybrid, or virtual in their published form, there are several important implications to consider when developing a successful strategic branding strategy and the ideal information product experience.

Brands are strategic, while visual identity guidelines are tactical. Brands are strategic messaging strategies, including the desired characteristics, messages, and themes intended to represent the information product. Visual identity guidelines function as the tactical blueprints that communicate brands through specified visual, spatial, and textual codes. These messages can be communicated both explicitly, through textual content, and implicitly, through design elements that support the overall brand experience. Information developers and designers implement these specific design conventions when creating information products, which are ideally informed by established information design principles practices, and techniques to ensure proper implementation. Together, strategic brands and visual identities align in their implemented messaging strategies to create cohesive information experiences for users.

Brands and visual identities should be carefully and iteratively planned. The development of brands and visual identities evolve through careful planning techniques, design strategies, and implementation methods. Brands are much more than simple slogans, but rather, when properly implemented by designers, they ultimately include the graphic, positional, semantic, and stylistic characteristics that communicate branded messages and themes. Brands and visual identities also evolve through multiple iterations and prototypes over the course of their development. Over time, a successful brand and its visual identity guidelines will likely change based on several factors, such as product updates, user expectations, and usability factors, rather than random novelty. While trends and novel approaches may temporarily boost the freshness of a product brand, trends inevitably fade over time and often require excessive resources to maintain. Also, more frequent and rapid changes are likely to create additional challenges and frustrations, particularly for loyal users. While brand evolution is often necessary throughout a product life cycle, the change in design, styles, and themes should be reenvisioned and revised based the discoveries, research, and user feedback that directly relate to the information product users and actual use. Throughout this iterative evolution, development strategies should demonstrate alignment between branded messages and product design features to ensure consistent messages and information experiences are communicated to its users.

Successful brands are supported by principles of visual and spatial thinking. Our perceptual and cognitive abilities are adaptive and help us learn and master new information environments, from our initial exposure and throughout their continuing evolution. As users interpret unique brands and experiences from information products, they think visually and spatially to help them interpret the various characteristics, content messages, functions, purposes, themes, and uses. Branding strategies and visual identity guidelines that emphasize important concepts, features, functions, and semantic relationships between the codes and messages used, will make the process of interpreting and learning new information product environments much easier for users. Therefore, information developers can integrate principles of visual and spatial thinking to design more intuitive environments that optimize usability.

Brands and visual identities should balance both consistency and creativity. While brands and visual identities inform the development of both content and design, they provide consistency through a variety of techniques, such as established conventions, style sheets, templates, and so forth. They also should accommodate enough flexibility to foster creative and innovative approaches to help products and brands evolve over their life cycle. While consistent features aid the findability and usability of information and their environments, creative approaches often add freshness, interest, and the variety that users often seek. However, too much creative change in a relatively short period of time, particularly in products frequently updated, can create information experience problems for users. For example, the case of the frequent, overdesigned information experience is all too familiar with a number of product apps and websites. Once users learn an information environment, their performance relies on this knowledge when they return for repeated use. When too many features change too rapidly, this can compromise the stability and perfor- . mance level users expect, and can create additional cognitive challenges and frustrations. While development teams may be eager to implement new features, technologies, and trends to improve information products, this can create an information experience that is unstable, unpredictable, and ultimately unusable, particularly for experienced users. Often, a more balanced approach between the implementation of creativity and consistency is necessary to maintain product stability. For example, consider how an information knowledge portal, such as Wikipedia, functions in terms of its overall information experience. When a user learns how to use tools in the environment, they learn how content is organized, navigated, searched, and used for various information seeking tasks. To keep the product fresh and on-trend, developers may decide to periodically update the content and environment, adding updated search tools, content topics, and related features. While keeping content fresh and the environment design consistent, users can rely on an information experience that is reliable, stable, and usable for performing regular searches for information on topics of interest. This, in turn, creates an information experience and overall brand, which facilitates use and supports brand and product loyalty among its users. And as a result, from the user's perspective, the product's functionality aligns with their overall perceived information experience. Strategic brands and visual identities that align with and support one another balance elements that users will recognize or expect, while introducing enough variation to keep users' interest over time. Successful information products are built upon similarly balanced strategies that ensure an information experience supports both innovation and stability throughout the information product life cycle.

Ultimately, successful strategic brands communicate the value of information products through carefully planned, developed, and implemented messages. Successful brands communicate those messages both explicitly, through subject matter content, and implicitly, through the various visual, spatial, and textual codes used in developing and designing information products and their environments. Branded messaging strategies should also align with the product's visual identity guidelines to ensure information is communicated both clearly and effectively between product and user. A well-branded information product incorporates both creativity and consistency, which rely on a combination of stable elements, as well as new or novel ones that support the information product and its use. Successful brands should also be informed by relevant user research, including their subject-matter knowledge, preferences, and how they think visually and spatially in information product environments to ensure a more successful information experience for them.

Chapter 6

Tactical Design

Information products and the experiences they convey are constructions, in part, of their users, content, and environment. The specific practices used in the design and development of these products are tactical in nature, encompassing both information design and user experience design principles, standards, and techniques. Consequently, these design and development tasks are typically grounded in practice, research, and theory within these disciplines. While information design focuses largely on style and presentation, user experience emphasizes access and use. Together, these two approaches, as related tactical practices, support information product developers in making informed choices in overall information product design and development. These tactics are also informed by both theory (including design theories and principles) and practice (including specific conventions or techniques). Tactical design practices must also consider users, including how they perceive, interpret, and comprehend visual, spatial, and textual elements within information products and environments. This includes how information product designs support various contexts, needs, purposes, and uses. Holistically, design tactics underscore the importance of design thinking to support products and experiences that are desirable, technologically feasible, and viable to both users and product developers (IDEO, 2024). And while a specific brand or visual identity may provide the blueprint or strategy, the use of specific design tactics implements the specific strategic themes throughout an information product.

Information design focuses on the presentational aspects of product design, including the use of individual characteristics and elements, such as images and visual styles, as well as the larger information environment itself, which includes the use of specific design layouts and positional techniques. Information design includes the design of various content properties, whether as static or interactive, monochromatic or colorful, informative or decorative, as well as their visual, spatial, and textual codes. It also includes the design of information graphics (such as charts or graphs), pictorial representations (both realistic and simulated), and visual brands (such as visual identities and holistic design concepts). Information design applies to all levels of design, from the use of fonts and text spacing to image and text alignment to overall page layout, and both at the micro (or element) level and the macro (or global) level of design. Information design work involves the application of design principles into specific practice. For example, when designing a product website, the use of the principle of contrast, may be applied using coloring, positioning, or shading techniques to add emphasis to various elements for effect. These applied practices might include the use of familiar shapes or patterns to create consistency or emphasis in the overall design. Ultimately, whatever collective features, practices, and principles are used, information design should demonstrate an overall unity that is on-brand with the product specifications.

User experience design focuses specifically on how we use information products, in terms of the overall accessibility, functionality, and usability. It also addresses the elements and functions that comprise the information environment, its interface, and how they support task performance and use. As such, user experience design also emphasizes the importance of user-driven decision making and heuristics that inform the development of information products. One particular focus is content accessibility to ensure users with specific access limitations, whether perceptual, cognitive, physical, technological, or others, have a consistent user experience, which adapts to their specific needs and preferences. This extends to the information product as a whole, as well as its individual functions and features. For example, when developing an information product wiki for technical support, the interface may include multiple tools for browsing and searching content, which must be flexible enough to accommodate different task flows. These tools may include a simple keyword search, conditional search, and a topic index. Depending on the purpose and preference, users may prefer a simple tool, such as a keyword search, for simple tasks, while using a more complex one, such as a conditional search, to sort and refine their searches iteratively. The design of these tools involves designing the information environment (including elements such as buttons, form fields, hyperlinks, and visual styles) as well as its functionality (what process users might follow, element functionality, and so forth). This example illustrates how information design and user experience design practices function together to create usable information environments and content. As the tactics of information experience, both information design and user experience design help information product developers create information products, content, and environments that best suit the intended users and uses.

How User Interaction and Thinking Inform Design Tactics

Ideally, our design tactics will consider the users' unique needs and preferences throughout the development process. While information design and user experience design are informed by established principles and practices, these practices must also incorporate the specific interactive contexts, including how users learn, perceive, think, and use information products. Information designs are created from a combination of visual, spatial, and textual codes, which suggests the importance of visual and spatial thinking in interpreting design codes, features, and environments. Within electronic information product environments, users are driven by intention, which suggests more active engagement, whether they are performing tasks, information searches, or simply browsing for entertainment. Users often have high expectations with regard to their information experiences in these product environments, such as informative content, intuitive features, organized layouts, and interactive properties to assist them with their information needs. To accommodate these expectations requires the successful integration of both information design practices (creating well-designed and presented content) and user experience design (designing interfaces with useful tools and features). Users expect content that encourages both push (passive experience) and pull (active experience) in their collective information experience. Therefore, incorporating methods and techniques that align with how users think visually and spatially in these information product environments is essential.

Visual-spatial thinking positions users as active thinkers, rather than passive bystanders, although our perception helps us switch easily between these roles in information environments. Our collective perceptual and cognitive behaviors help us focus on elements that help solve problems, form concepts, and conceptualize the whole (Arnheim, 1997). Both user and content interact dynamically within information environments. Sometimes, content dominates the experience through its presentation, while in others, the user dominates the exchange through their interactive behaviors and choices. While the information exchange between user and system may vacillate between push and pull actions, almost all information products, whether hybrid or electronic, require varying levels of user interaction and engagement to be optimally successful. Interactive environments typically require more deliberate than passive ones, such as simply turning pages or clicking directional arrows to control the flow of information. For example, in virtually any electronic information product, users must interpret and learn the function of basic navigation systems, in addition to any other purposes they may have. It requires higher order visual and spatial thinking to interpret navigational tools, interactive forms, multimedia content, information structures, and even virtual environments (Johnson-Sheehan & Baehr, 2001). This is particularly true with information products such as knowledge bases, virtual simulations, websites, and wikis, which can incorporate highly complex interfaces and functions. As a result, our design practices have evolved to incorporate more sophisticated and varied visual and spatial codes to help users with access, performance, and use. While textual design was prominent in the design of print-based information products, such as manuals and reports, hybrid and electronic information products have more complex visual and spatial elements that fundamentally change the way users think and experience information. Consequently, our corresponding design principles and practices are highly influenced by these changes, which include how users think both visually and spatially in these unique information environments.

Principles of visual and spatial thinking, informed by Gestalt design theory, can also be used to describe how users interact with various visual, spatial, and textual elements in hybrid and electronic information environments (Baehr, 2007). Visual and spatial thinking offers designers many useful applied guidelines to help develop information products and environments that align closely with users' perceptual and cognitive processes. As an example, one of these guidelines suggests that information designers should assume that users are actively trying to solve problems in information environments (Johnson-Sheehan & Baehr, 2001). These problems can be as simple as learning or mastering new information or processes, or they can be task-based, such as registering for service or using an information search tool. To assist users with this problem-solving mindset, visual elements can support these kinds of user interactions, whether they provide assistive illustration, concept formation, emphasis, highlighting, performance enhancements, or simplify messages.

Within web-based information products, other visual-spatial design techniques might include using framed space to anchor contextual information, structuring navigation tools to highlight future paths, incorporating design metaphors to create three-dimensional (virtual) spaces, using icons and images to suggest concepts or terms, and designing a flexible and dynamic interface to optimize use (Johnson-Sheehan & Baehr, 2001). As another applied practice, the use of simple, recognizable, or repeated shapes places less cognitive demand on users, enabling them to comprehend information environments more easily. The use of simple and consistent shapes throughout a design supports concept formation and efficiency (Johnson-Sheehan & Baehr, 2001). Conversely, using unfamiliar shapes might cause added effort or confusion for users, where providing appropriate context, such as supporting descriptions, labels, or alternate forms of content, supports user comprehension. Sometimes, simple pairing of visual and textual elements can provide the contextual details that users need. These pairs can suggest relatedness or similarity in function or use. Since basic concept formation is part of visual and spatial thinking, information designers should pay particular attention to how elements are positioned, related, and repeated throughout an information product, as well as how they convey specific semantic meanings for users.

Within information products, users often plan different paths when browsing and searching, and it is important that systems support their instinctive visual-spatial thinking and behavioral patterns that optimize their information experience. One visual-spatial design technique includes designing flexible navigational systems that allow users to follow associative and semantic paths through an information product based on related concepts or functions. Integrating contextual cues, such as breadcrumb links, descriptive headings and hyperlinks, enhanced styling for emphasis, or even performance tool tips, can add semantic value to their navigation experience. Collectively, these applied guidelines demonstrate how information design and user experience design practices align with visual and spatial thinking and function as overlapping design practices, particularly in the development of complex information environments, such as websites.

The Tactics of Information Design

As part our specific design practices and tactics, information design includes the development and use of supporting visual, spatial, and textual codes and related content, often in multiple combinations or formats, to communicate simple and complex information. Put simply, information design may use images or words, print or digital forms, and applied practices to clarify complex information, which is tailored for specific purposes and users (Black et al., 2017). It also involves the use of presentation techniques for both content elements (figures) and the environment (ground) of an information product or design as a holistic composition. Information design incorporates both principles, which are broad theoretical guidelines, and conventions, or specific techniques which are used to create information products (Baehr, 2007). Specific principles and conventions used to design an information product can include a wide range of properties, such as use of color, depth, emphasis, layer, position, shading, spacing, style, and others. Ultimately the designer must make informed choices using design principles and tactics in ways that best fit their purposes, contexts, and intended product uses. These choices should also be informed by what we know of how users perceive, understand, and interact with these information products.

Design principles and tactics are typically informed by foundational design theories and approaches based on factors such as behavior, cognition, culture, perception, semiotics, and others. Gestalt theory, discussed in a previous chapter, provides several foundational core principles that inform many design practices and techniques that are widely used. While information design textbooks in circulation often use different principles and terms, collectively they center around design concepts that emphasize alignment, contrast, consistency, positioning, and repetition. Many of these design principles incorporate theories of perception and cognition, which align well with how users perceive, think, and interpret elements of design in a wide range of information products and environments.

Information designs also function as a combination of codes, including visual (or graphic), spatial (or positional), and textual (or linguistic) elements, that function both independently and collectively at different levels within a document design, whether at the element, line, page, or document level (Kostelnick, 1996). For example, consider a branded set of visual identity guidelines, which may include specific colors and schema (visual), fonts, slogans, titles, headings (textual), and positioning and alignment specifications (spatial). Each of these design properties can have function and meaning at multiple levels within an information product, whether at the element level (specific to a content chunk or element), page level (specific to all content on a single page), or global document level (applied across multiple pages). Cascading Style Sheets (CSS) is a scripting language commonly used to create design conventions and specification for web-based content and can apply design properties and values at all levels of a document. These style sheets provide many of the design conventions for use of backgrounds, colors, basic page layouts, and so forth, while local style sheets may provide page or item specific style sheet declarations, which govern the position and styles of text, visuals, and even use of negative space. Many web development platforms incorporate the use of multiple overlapping CSS style sheets to customize and implement the information design conventions and specifications used for a particular information product.

The Core Principles and Features of Information Design

Design principles provide broad guidelines, based in design theory, that can help information designers make informed choices in their work. While design principles provide broad guidelines that can be applied to any information design work, they are often interpreted into different practices, depending on the unique contexts, purposes, and users, as well as the expertise of the designer. While design principles are strategic, their application as design conventions are tactical. These principles inform the development of specific design conventions, style sheets, and unique specifications of an information product. Design conventions emerge from the interpretation of principles into specific, discrete practices and properties that specify the use of visual, spatial, and textual codes throughout an information design. Design conventions function as specific applications of information design principles, tailored to the unique brands, purposes, and contexts of information products. When tailored appropriately, they help form a more cohesive, consistent, and unified design concept. Design conventions provide the specific characteristics and standards for style sheets, style guides, templates, and visual identities. Design conventions specify the properties and values for a wide range of design factors, such as the use of styles, shapes, position, grouping, or other elements. While design conventions may be consistent from one product to the next, such as when a particular brand, standard, or set of expectations must be maintained, they will undoubtedly vary from one information product to the next.

Five core tactical principles of information design that align with both Gestalt theory and visual thinking and can be readily applied in design work include the following: conceptual, consistent, contrastive, positional, and relational. These core principles also align with the principles of Gestalt theory, which include the concepts of closure, continuation, figure/ground, proximity, similarity, and wholeness. While the specific names of design principles may vary in information design references and textbooks, these core principles can be considered foundational concepts, which holistically represent coherent information design practice. To illustrate, several examples will be explored that relate to each of the five core principles and how they can be used to aid comprehension, performance, and usability in an information design.

Conceptual design tactics focus on the communicative and semantic meanings conveyed in design elements, whether through a single element, a group of elements, or in the entirety of the environment. When interpreting elements in our visual field, we may discern specific conceptual meaning from colors, icons, images, patterns, shapes, and symbols used independently or in combinations. These concepts may suggest characteristics, functions, messages, metaphors, or themes, which support the overall visual brand or design theme. Our ability to form concepts from design elements relates to the Gestalt principle of wholeness, which suggests our perception helps us form complete ideas or concepts from the disparate visual, spatial, and textual codes (Kohler, 1947). Using specific tactics that convey concepts, such as function or use, can support comprehension of specific information product features. In turn these concepts can support information findability, navigation, task performance, and overall usability within an information environment. For example, the use of icons in a mobile device's software interface may suggest specific functions, such as a camera icon for taking photos, an envelope for accessing email, or three condensely grouped horizontal lines, also colloquially known as a hamburger stack, to represent a pop-up menu of choices, functions, of settings (see fig. 6.1). Users may transfer their knowledge of these shapes or concepts from other information products and contexts; however, if the intended meaning is different from these expectations, it may be necessary to provide contextual clues (such as text labels) to aid comprehension.

Figure 6.1. Example of conceptual design. Conceptual design elements can suggest specific functions, meanings, or messages when they are used in ways users expect or with specific contextual clues. *Source*: Created by the author.

Conceptual design techniques can also communicate specific characteristics or functions through the use of various images, symbols, and styles, which users may learn and recognize. Because concepts are not always universally recognized, it may be important to include other contextual cues, such as using text labels or descriptors, which help convey the intended function or meaning. Similarly, pairing a directional arrow symbol with a textual descriptor such as "advance" or "next" might suggest how users can navigate through a series of pages or screens, which otherwise may or may not necessarily be understood without such contextual messages. Conversely, using abstract, unfamiliar, unusual shapes, or misaligned visual-textual pairs may impede overall concept formation (or comprehension), resulting in user confusion, frustration, or performance issues. For example, an airplane icon used in a website navigation toolbar, without a label, may suggest multiple concepts or meanings depending on its use. This symbol can have different meanings when used on a printed safety information card, on an airline booking website, or in a live simulation. When using design elements that support concept formation, it is important to provide appropriate context, particularly for elements that may have unexpected or multiple meanings, to ensure clear communicative messages. Pairing abstract elements with concrete ones, such as icons with textual descriptors, can support overall comprehension and recognition. And since basic concept formation is a key perceptual ability for users, it also serves as an important design consideration that supports performance, use, and the overall information experience within a design environment.

Consistent design tactics emphasize the importance of repeated and similar design elements, which establishes recognizable patterns, styles, and themes throughout a design concept. Repeated features can include borders, colors, patterns, spacing, styles, and, in essence, nearly every aspect of an information design. The Gestalt principle of continuation relates closely to consistency, in that the repeated use design elements reinforce meanings and patterns that are recognized by users (Koffka, 1935). Consistency also underscores the importance of design elements that are prominently repeated to reinforce specific patterns throughout an information design. For example, visual identity guidelines, used in visual branding, use similar design elements in consistent or repeated ways, such as in the use of specific logos, images, colors, font styles, and other elements used to reinforce the particular design brand. Consistency can reinforce specific themes through their repeated use, communicating the grammar and syntax of the visual language, of sorts, that is used in a design. Style sheets, which are commonly used to formally establish these conventions and standards within projects and documents, demonstrate the importance and implementation of consistent design styles and techniques. Design style sheets and templates are the blueprints of design conventions used in creating holistic design concepts that follow a specific brand, identity, or theme. In web-based environments, Cascading Style Sheets (CSS) is a scripting language frequently used to create consistent style declarations and global style sheets, which enforce consistent design standards in electronically based information products. Often, style sheet declarations can be used to implement consistency at multiple different document levels, whether at the unit, line, section, page, or global level. And in some cases, multiple style sheets work together to apply consistent standards throughout an information product, or even an entire product family. Accordingly, consistent design strategies can support stability and cohesiveness in an overall visual identity or brand. Whether at a global level or component level, such as a simple functional toolbar, consistent style supports comprehension and performance. For example, the use of similar borders, colors, shapes, sizes, and spacing of icons used in a simple application cluster, which we might see in a mobile phone interface, can communicate the function of each individual item, such as a list of upcoming flights or trips, bookmarked stock trends, email messaging app, or summary of recent news headlines (see fig. 6.2).

Consistent design techniques can include repeated patterns and styles, which reinforce comprehension and communicate cohesiveness,

Figure 6.2. Example of consistent design. The use of consistent design elements support stability in a design, which aids comprehension and performance, such as application icons that incorporate similar borders, labels, styles, and grouping or spacing techniques within a product interface. *Source*: Created by the author.

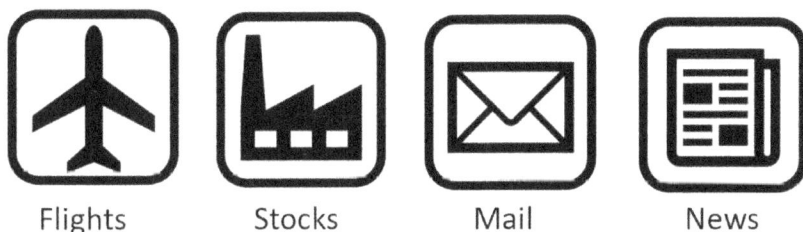

| Flights | Stocks | Mail | News |

such as a simple set of functional buttons used in an application interface. Consistent design tactics can help users perceive these repeated objects as recognizable patterns throughout an entire information product. For example, using consistent style declarations (font face, font size, text color, and indentation) for a singular textual heading can suggest its importance, location, or position within the overall information structure. In turn, this can also reinforce particular user behaviors and expectations, when they learn the meaning of these patterns, which are used consistently throughout an information design or product interface. While consistency is often desired, in some cases, it may not be useful—particularly when it is necessary to highlight difference or emphasis. Users naturally notice differences in visual presentation, whether subtle or distinct, and often perceive them as having different meanings or relationships to other elements present in the same environment (Arnheim, 1997). When creating dissonance or emphasis in the use of design elements, nevertheless, it may be important to use consistent styles or techniques in ways that support these differences, to aid comprehension. However, when design elements are used inconsistently or unintentionally without proper supporting context, they can create disparity where perhaps none was intended. Something as simple as a random use of colors, shading, or even image quality can cause users cognitive difficulty in interpreting the proper meaning of these design elements.

Contrastive design tactics emphasize the visual distinctions between two or more elements when comparing them, typically in a single space or as a whole unit. Contrastive techniques focus largely on the use of visual codes to create emphasis or communicate the specific value of elements within a design, such as a brightly colored border or darkly shaded background. Despite the importance of visual codes in any design, often the supporting spatial and textual codes may provide additional contrastive features to communicate the desired effect or message. These codes may include the stylistic choices used to create contrast, effect, or emphasis within a design, which affects how elements are noticed. Arnheim (1997) suggests we naturally perceive visual elements in a visual hierarchy of sorts, which helps us prioritize elements by order, semantics, or usefulness. Users will typically notice distinct visual elements before all others, which may include animated or high contrast design elements, while more subtle elements are noticed and prioritized lower in our visual hierarchy. This hierarchy functions on a continuum, of sorts, from a high to a low level of distinctiveness. In turn, our perceived visual hierarchy may represent the perceptual and cognitive priority we assign to each element we wish to examine in greater detail. While we may naturally (or instinctively) respond in similar ways to high and low contrastive elements in our visual field, our experience can also affect our responsive priorities. And despite the level of visual distinctiveness we attribute to each element, we can override our natural response based on our previous experiences with similar elements (Baehr, 2007). For example, users learn to ignore animated visual advertisements in a website if they deem those messages to be bothersome or irrelevant to their information needs, based on prior experiences. While various elements compete for our attention within complex environments, our combined perceptive and cognitive abilities help us compare and discern elements as components within the design, to help us better understand the distinctive features between them.

Other specific contrastive design techniques typically include the use of color, layering, shading, typography, or other stylistic coding. Using different font faces and font sizes for heading levels can support understanding of information hierarchies, such as differences in categories, which may improve scannability or quick comprehension. Supporting spatial and textual codes can also emphasize other differences in information levels within a document, such as indenting or shading techniques. Color can also be used to communicate the specific meaning of independent data sets within an information graphic, through contrastive techniques. When multiple stylistic codes are used together, such as in the developing of a visual identity, contrastive techniques may be used as part of this visual branding to emphasize specific patterns, shapes, or themes. Contrastive design techniques relate to the Gestalt concept of figure/ground, which suggests that visual compositions have a both foreground (the figure) and background (the ground), underscoring the importance of contrasting features between two or more visual elements in the same space (Koffka, 1935). For example, when viewing graphic art selections from an art gallery's website collection, each image may be featured in the foreground (as the figure) with a bordered frame hanging on a plain white wall space (as ground elements). If the background incorporates neutral colors or simple patterns, each image (as the figure), may be enhanced or emphasized, functioning as a visual focal point of our attention. Each combination of figure and ground elements establishes its own communicative structure as a functional unit. Subsequently, each iconic image presented in foreground (or figure) provides a coherent focal point, while a simple black bordered frame and negative spacing (or ground) function as the supporting visual backdrop or environment (see fig. 6.3). Collectively, each figure/ground pair provides a simple, but effective combination of characteristics, which improves the overall viewing experience.

Contrastive techniques are not always used effectively. Compositions that improperly use contrastive techniques may create designs that fail to emphasize key features, obscure important details, or impede comprehension of the design and its overall message. Contrast can integrate both low (blending) and high (emphasis) techniques together; however, using too much or too little contrast can cause other design problems (Johnson-Sheehan, 2024). Low contrast blending techniques establish a more subtle relationship between figure and ground elements when one or more layers is supported by the others. A single figure, such as an icon, may be supported

Figure 6.3. Example of contrastive design. Contrastive design techniques can be used to create emphasis and a focal point between figure (an object) and ground (its framing or spacing), to draw attention to specific features, elements, images, or messages. *Source*: Created by the author.

by multiple grounding elements, such as a watermark and lightly colored background. In fact, low contrast blending may be perceived as somewhat invisible, since these details may be barely noticeable or even ignored in favor of a highly contrasted figure. Conversely, high contrast emphasis can enhance a figure's semantic meaning or value, which may promote their importance and noticeability within our perceived visual hierarchy. For example, in websites, mouseover effects, which are triggered when users place their mouse pointer over an object, often alter the visual properties the object, indicating it has been clicked or hovered over. This change in contrast or emphasis creates a new focal point to gain our attention. While this contrast can be established through the simple use of figure/ground, it can also convey similarity or difference. When layering multiple visual elements in the same space, whether figures or grounds, these techniques are examples of how to create emphasis, focal points, or separation based on variations in contrast and styling techniques used. Contrastive design techniques also support the overall comprehension of visual codes and messages present in an information environment through this conceptual pairing of figure and ground elements.

Positional design tactics focus on the importance of spatial codes, which include the placement and spacing of individual elements relative to others. Positional design techniques may also incorporate visual codes, such as the use of borders, lines, patterns, shading, and even some textual codes, such as typographic features. Positional tactics are closely related to the Gestalt principle of proximity, which suggests that design elements share similar or relational aspects based on their closeness, distance, or location, which affects our overall perception and understanding of these elements (Kohler, 1947). Like other principles, positioning or spacing of elements can create dissonance, emphasis, prominence, or relatedness, which suggests the importance of semantics in design. Element placement, when in predictable or noticeable locations on pages or screens, can convey specific repeated semantic functions or messages to users, such as footers, headers, or navigational aids. Position can also signal disparateness, using negative spacing between elements to create specific design grids, layouts, or organizational patterns. These patterns may suggest different information hierarchies, such as in a table of contents or site map, or even suggest comparisons, such as equally spaced objects placed in a row or column. For example, an equally spaced grid of similar product images and descriptions, such as audio headsets, or equally spaced icons in a navigation toolbar, suggest similarity or disparity between individual elements. A basic navigation toolbar may combine multiple visual, spatial, and textual elements, such as the use of icons, text labels, pipe symbols, and negative spacing, which in combination suggest a series of separate functions (see fig. 6.4). While each unique pair has its own independent function, the collective group provides the full range of tools available for use. These positional (and consistent) design grouping techniques can also support comprehension and performance in the interface in which they are presented for use.

While positional design tactics emphasize the use of spatial techniques, often supporting visual and textual codes can support page or screen layouts. Positional tactics often communicate visual-spatial structure between elements, or within larger organizational patterns (or grids) as well. For example, pairing visual and textual elements, such as an icon and a text descriptor in close proximity can communicate shared meaning, where one element enhances or supports the other. In a larger sense, grouping elements on a page or screen using space, such as in a website home page layout, may help users discern different information groups of functions, such as header, navigation toolbar, content window, search feature, footer, and so forth. Positional design techniques can be combined with other tactics, such as conceptual or relational techniques, where close placement of elements appears to frame or group elements with shared conceptual meaning. Conversely, positioning or spacing techniques may also emphasize difference between a set of elements if users perceive sufficient distance or space between elements, which might suggest a lack of relatedness between them.

The position of design elements can communicate both independent and collective meanings within a design, as well. Individual charts and graphs within a complex data presentation can have their own independent meaning while also suggesting a collective one when individual elements

Figure 6.4. Example of positional design. Positional design techniques can suggest both similarity (a collection of navigation options) and difference (independent functions), through the use of spacing and visual styling. *Source*: Created by the author.

are placed within close proximity. To illustrate, magazine infographics often show multiple statistical views of data, such as a bar graph, line graph, pie chart, and others, on a single page. Together, these data displays form a singular complex group to illustrate a single concept. Demographic data used in such displays might depict population statistics based on age group, education level, ethnicity, gender, geographic location, and so forth. Multiple information graphics based on geographic location might be positioned (or grouped) together on the same display for added complexity. Within a single information design, element positioning might also suggest equal or subordinate relationships between individual objects placed in the same region, or even on the same page. Positional design techniques may also affect the balance or visual weighting of elements on a page, which can be used to create both asymmetrical and symmetrical page designs based on the use of space (Johnson-Sheehan, 2024). Website navigation toolbars sometimes use icons to depict specific functions, such as search (hour-glass), help (question mark), information (a lowercase letter *i* enclosed in a circle), or hyperlink for the home page (house symbol). Equally spaced icons may suggest a series of choices and separate functions, which can be discerned from its symbol (visual), label (textual), and position (spacing). This illustrates how one design principle (positional) works symbiotically with another (relational). Similarly, positional design may also rely on consistent techniques to convey stability within an information design.

Relational design tactics emphasize the importance of similarity and difference between elements in an information design. Relational elements form conceptual wholes, whether by closeness, distance, distinctiveness, similarity, or subtlety. When individual elements are paired together, they create configurations, which are perceived to have shared characteristics. Relational design tactics relate to the Gestalt principles of closure (or connectedness) and similarity (or likeness) of elements (Koffka, 1935). Design elements with the same shape or color, such as functions in a graphic design software program toolbar, may share relational similar-ities—each representing a major category, such as drawing tools, while each independent tool has its own unique function. Within each grouped toolbar, different functions may be perceived as having semantic similarities, such as different pen or pencil drawing tools. As another example of how individual elements can suggest relatedness, consider a scene with a car parked next to a streetlamp in front of an office building (see fig. 6.5). We perceive each individual image or shape as its own complete form; however, collectively, their grouping may have a shared meaning—such as

Figure 6.5. Example of relational design. Relational design techniques can be used to pair elements within the same space to communicate relatedness, semantic connections, and a conceptual whole, or holistic meaning. *Source*: Created by the author.

a simple urban landscape. Despite the fact that one object may partially obscure another, such as the car appearing to be parked behind one of the streetlamps, with enough details present, our sense of closure helps us interpret any partially completed shape (such as the partially obscured car) as a complete whole. However, if there are insufficient details present, such as the case of a second car placed behind the building, we may be unsuccessful in perceiving its meaning or relation to other objects present. Regardless, our visual perception helps us discern each individual shape or image, complete or incomplete, as well as its relationship to others, which helps us form a holistic understanding of the urban scene.

Relational design techniques, through such configurations, can convey semantic connections by using familiar shapes, paired elements, shading, spacing, and other visual styles. While proximity and grouping techniques can support relatedness between elements in a design, the use of other techniques can also support these messages. Pairing visual images and text labels can also communicate relatedness, since the use of these visual-textual pairs often communicates messages more directly and easily to help avoid misunderstandings in their individual usage. In such a way, relational design tactics can convey important semantic messages within a design, which are communicated through various visual, spatial, or textual codes present. Relational design techniques can also support holistic understanding within an information design. To further illustrate, consider the multiple design specifications that comprise a design aesthetic, such as minimalism, or even a familiar brand. While each aesthetic or brand incorporates a set of different design declarations, such as the use of subtle contrasting colors or sans serif font faces, collectively these techniques function as a coherent, holistic design concept. Each individual element within the design supports the brand, which uses a set of related design techniques that best communicate the intended effect. Ideally, the result is a holistic design concept with individually related design declarations that successfully communicate a particular aesthetic or brand.

While the five core principles of tactical information design encompass a wide range of design characteristics, they are not mutually exclusive, and, in fact, often overlap in their application and use. For example, positional design techniques focus on spatial relationships, such as an object's proximity to others, and, as a result, they may communicate different relational or semantic messages based on position or placement. When two objects are placed in close proximity or on top of one another, it might suggest a close relationship or similarity as points of comparison. This example illustrates how positional and relational tactics overlap in conveying a specific message. While design principles offer no absolute rules for usage, their theoretical foundations and potential applications provide guidelines for developing appropriate conventions and techniques used in information design and development. Additionally, individual design aesthetics can often influence information designs, which may reflect individual preferences, established themes, or even commonly used design concepts. For example, an established design concept might be characterized as contemporary, minimalist, modern, or traditional. Consequently, each concept has its own unique set of established design characteristics and conventions. To illustrate, an information product with a minimalist style might incorporate the use of mostly subtle, neutral colors, and plain style fonts, minimize the use of highly contrastive visual images, and favor the use of negative space to present clutter-free, organized design spaces. Design aesthetics can also be guided more by individual preferences that may have an appeal with some users, but not so much with others. For example, design templates often found in document design software products are typically based on an established theme, style, or particular aesthetic. Templates may represent a particular information product brand, including variations such as a business presentation, financial report, or technical product description. Modifying template designs for a particular user group or product is usually an improvement over using a stock template, but may require specific product or user research. However, due to the wide range of design aesthetics and individual preferences for a particular scheme, it is often difficult to create designs that will appeal in the same way to all users. Regardless of the particular aesthetic used, an information design should also be informed by established principles and tactics to ensure successful application.

The Tactics of User Experience Design

User experience design focuses specifically on the iterations and uses of information product, informed by what is known about users and their aptitudes, habits, and preferences. User experience design involves the creation and synchronization of elements and features that affect both the behaviors and perceptions of an information product's intended users (Unger & Chandler, 2012). Through the development process, users can inform the creative practices of designing user-centered products. User analytics can be collected from individuals or groups through automated or manual processes and in physical or virtual spaces. Depending on both purpose and usefulness, user analytic data can be collected throughout the information product development process, informing prototypes and product iterations to synchronize both users and intended information experiences. While more user-centered approaches incorporate user analytics throughout the entire development process, other less process mature approaches may only do so sporadically, or not at all.

User experience design focuses specifically on the usability of information products, including approaches that focus on user appropriation and use, such as interaction design and user-centered design. User experience design encompasses a wide range of user modalities, including aural, haptic, and visual aspects of information products, which may be both physical (keyboards, mice, screens) and digital (interface elements, multimedia content, navigation tools). While user experience design focuses primarily on design usability, it remains a critical part of creating information experiences well suited for product users. Ideally, these experiences are also equally informed by user perception and cognition, how users behave and think visually and spatially, and the unique brand and identity of information products that create universal appeal.

Theoretical Approaches to User Experience

User experience design is informed by many different approaches or ways of thinking about how information products are iteratively developed, prototyped, and tested for their intended users and uses. User experience design is also an integrative, holistic process that attempts to align environment, functionality, and user. Garrett (2010) conceives user experience as layers of concrete and abstract elements, integrating them into a cohesive and highly usable information product. The concrete elements include the applied practices of information design, interface design, and navigation design, which govern how specific content elements and features are developed for use. The abstract elements focus more on user and project concerns, such as the overall objectives, user needs, content requirements, and functional specifications. Two applied practices, information architecture and interaction design, are classified as hybrid elements, which suggests a mix of both abstract and concrete practices in their development. Holistically, these abstract and concrete practices describe how user experience design integrates the unique specifications of information products with all aspects of user experience, including content development, design, interaction, navigation, and structure. These elements also suggest the importance of three primary contexts in which users experience information: perceptual, cognitive, and rhetorical. From a perceptual standpoint, users actively seek out visual objects, solve problems, attempt to identify concepts, and comprehend contexts as part of the user experience (Arnheim, 1997). Cognitively, users make meaning, interpret structures, contribute content (narrative and dialogic engagement), acquire feedback, and make decisions. And rhetorically, users possess unique constraints, motivations, and purposes that guide their interactions with information products and their environments.

One design approach conceives user experience as a series of input, processing, and output actions, which can be cyclical, depending on the action being performed or intent of either the user or the system (Crawford, 2003). This algorithmic approach involves a rather simplistic, cyclical process, which loops through multiple iterations of user inputs, system processes, and feedback output from system to user. This approach focuses on the importance of content flow between system and user, creating a seamless information experience. Input is a form of listening, by the system, which includes looking for what to ask, collect, or permit from user interactions with the content environment. Users provide input through a wide range of actions, including clicking on hyperlinks, filling in form fields, keyword searching, navigating, and using various forms of interactive media. For example, when using a grocery delivery app, everything you click on is a form of input and much of it is tracked and handled (or processed) in some form. Whether you're clicking on particular items to view, purchase, browse, or comment, each of these actions is a form of input, and the system is listening, or monitoring these input responses. Past purchases might even be stored on a frequently bought list, to help users with their next grocery shopping experience. From a developer's perspective, determining how input is processed can have a direct effect on both the functionality and user experience.

Processing is a form of system thinking that is automated and often based on conditional statements and variables (Crawford, 2003). Processing involves how the system handles the user's input and often dictates the system's output and response. Both user input and behavior can vary widely, including what tools are used, which navigation pathways are explored, how much time is spent on a page, what mistakes or errors occur, and other conditions. System processing might include the use of algorithms, counters, databases, functions, queries, variables, and other constructs. Processing can occur for almost any system task, whether it is to calculate, customize, retrieve, or store information. Processing can also be restricted based on certain conditions, such as the time, location, or other factors independent of the user. For example, in a software program, installation of updates can be scheduled by user preferences for days or times, or they can be performed automatically by the system as they become available. This type of conditional processing can be problematic, when not used properly—allowing a system to handle information without properly applied user context (or input) can create disruption or frustration in the user experience.

Output provides the system responses, which is the end result of both input and processing that can be prompted by both user and system, depending on the conditions and desired responses. System output might include changes in the content, design, interface, and interactive properties, which can be obvious or even hidden. Output can also be considered as a form of feedback provided by the system for users, which might include messages of successful task performance or transactions. For example, when a user completes a registration form on a website, they typically provide data in various forms, through form fields such as checkboxes, radio buttons, select boxes, or text boxes. The data (or input) they provide might include their name, address, email, phone number, and other details pertinent to their registration choices. After the system processes the input data, it can provide its feedback (or output), such as confirmation details and other useful content. System output can also be hidden, such as storing specific data on user preferences for later use. This data might be later used in generating analytics reports or recalled when users return to the site to help them expedite performance. Accordingly, this output creates engagement between the system and user, enabling the creation of more customized experiences. And whether the output is a direct result of user input or other conditions, it can demonstrate responsiveness, allowing information products to be more interactive and purposeful.

While the input-processing-output approach may seem simplistic in concept, as illustrated, it can include a series of complex tasks and decision-making points in its actual application. As a result, this approach has some important implications to consider when creating optimal user experiences for information products. Aligning input methods with user experience goals is essential to overall information product usability and information findability. Systems that collect input that is unnecessary or irrelevant may waste the user's time and effort, resulting in a poorly perceived experience. This pitfall may be the result of automation or design. For example, when using a web-based content management system, developers should know what data is collected and how it is handled, or processed, by the system. In many cases, developers must thoroughly test and customize these options to ensure an optimal user experience. While this may require additional technical expertise or resources, it is nevertheless an important consideration to avoid allowing the system to handle information without appropriate customization or context, which considers both users and uses of the system. Also, providing meaningful feedback (or output) will contribute to a more positive user experience, particularly when it is customized to their input and unique preferences. Output, which is counterproductive to user performance, or generic in nature, should generally be avoided. For example, many online shopping websites provide suggested products that are popular or similar to prior purchases. Sometimes, these recommendations can be perceived as random suggestions that lack proper context sensitivity, which can be off-putting to users. As result, systems should prioritize useful feedback by asking the right questions, selecting appropriate input types, and aligning suggestions more specifically with user needs.

Another user experience design model focuses on task-based design, which involves an iterative series of prototyping and testing in developing an information product. Unlike the algorithmic design model, this approach focuses on the development as a function of the product, rather than the flow of content. Unger and Chandler (2012) characterize this approach as involving information architecture, interaction design, and user research as the primary areas of emphasis. Information architecture focuses on developing an overall structure and organizational patterns for an information product, optimized for use. Interaction design focuses on developing interactive patterns and paths, which allow users to perform critical processes and functions with an information product. User research focuses on collecting information about users and their uses of an information product, including developing user profiles, use scenarios, and task flows, which inform all aspects of the user experience.

Specific tasks related to these three areas of emphasis include developing site maps (overall site structure), task flows (procedural or process-oriented activity paths), annotated wireframes (interface layouts), and prototype information products (working models), which help create optimal user experiences (Unger & Chandler, 2012). Site maps provide a framework, often composed of visual, textual, and spatial codes, that represents an overall structural map of how information is organized. They provide high-level views of the major content areas, functions, and sections in an information product. When site maps integrate hyperlinks, they can also function as navigational tools, which helps users with positional awareness and an overall understanding of information structure. On the back-end, site maps function as useful patterns for organizing an information product and its contents, while on the front-end, they function as navigational aids and tools for users. For example, an online clothing retailer may categorize products in its website based on the type of garment, such as casual, formal, semiformal, and sportswear. In turn, each category may include subcategories, such as shoes, shorts, T-shirts, and sunglasses for the casual category. Organizing the site map using these categories and subcategories can support information findability as well as navigation of the site's contents. Task flows project activity paths or processes that attempt to map the user experience when performing certain procedures or tasks in an information product (Unger & Chandler, 2012). Task flows can be communicated through a series of steps users follow to complete a process, such as registering for an online class or webinar. For example, think of how a shopping cart might function as a task flow when purchasing a product through an online retailer's website. There might be multiple steps in completing the process, which are mapped out as individual tasks following a linear sequence. This might include tasks such as selecting products, reviewing the order, entering shipping or billing addresses, entering payment information and discount codes, and completing the purchase. When this overall process is mapped out as an activity path, this helps create a logical and seamless user experience. Task flows can also be developed and improved iteratively through user feedback and testing to optimize performance and use.

Wireframes are working sketches or skeleton frameworks of an information product's interface layouts and task flows, which can include navigation tools, content pages, headings, spatial separators, and other stylistic elements that can be used for iterative development and testing (Unger & Chandler, 2012). In some cases, developing alternate versions of wireframes for the same information product can be useful for testing different design concepts or interface patterns for their usability in a working or situational context. In their most basic sense, wireframes are the precursor to fully developed product prototypes. Prototypes function as working models of an information product, which can vary in their functionality, from low to high fidelity, in terms of their level of detail and development. Prototypes are more highly developed versions of wireframes, providing actual content and working features for evaluation and testing. They can be used to test one or more features, such as an advanced navigation search page or interactive registration process, or encompass the whole product environment, which can be tested in a proper working context. While task-based user experience models emphasize the importance of iterative development of an information product, they should be equally informed by user research as well as iterative testing cycles to ensure an optimal user experience.

Another user experience design approach, user-centered design, focuses on the importance of integrating user preferences and habits throughout the information product development life cycle. Johnson (1998) describes variations in how user research informs product development as three different approaches: user-centered, user-friendly, and system-centered design. These approaches focus on varying levels of emphasis on both user and system. Some product designs are created independent of user research, where the product developer envisions a system with its own unique features and language, which may not be intuitive or easily understood by users. This approach, known as system-centered design, focuses heavily or exclusively on the product and its uses, independent of users. A system-centered design approach focuses on the development team's concept of an overall product experience. In some cases, this may be a beneficial approach with specific advantages. For example, Hypertext Markup Language (HTML) was created as the primary markup language used to mark the structural and semantic features of content so it can be properly interpreted by a web browser. To become proficient in HTML, coders must learn the intricacies of the language, including the rules, syntax, and tag sets. Many coding languages, such as HTML (and its predecessor SGML), evolved from other languages, often independent from the preferences of its users. This system-centered approach, while largely ignoring user preferences, provides consistency and stability for the language and its use in developing a wide range of information products. Products that are developed independent of the user may, however, require more effort to learn, supporting documents, help support, training, or workarounds for users to master the information product environment and its contents. Product or systems that focus on the system, and, to a lesser extent, the user, are characterized as a user-friendly experience design (Johnson, 1998). User-friendly design applies practices that cater to the user, by improving aspects such as access, aesthetics, design, lay-out, navigation, and use, but typically occur after the initial phases of product development, such as at the wireframing or prototyping stages. This approach may be used when there is limited expertise, resources, or process maturity for a particular information product. And finally, a fully realized user-centered design integrates user research throughout the entire product development process. In user-centered experience design, user profiles, use scenarios, and user preferences inform the planning, design, and iterative development of information products, even beyond their initial versions. User-centered design processes resemble higher level process maturity flows, where usability, user research, and product innovation are all hallmarks of product development (Hackos, 2007).

Invariably, information product developers incorporate a wide range of approaches and techniques for creating successful information products. Over time, the development process may change or mature into one that integrates user research, use analytics, or other techniques designed to improve the overall user experience. Within each of these frameworks, design practices may vary based on products, purposes, and contexts. For example, many system-centered approaches have been used to create new interfaces and products that have their own unique features that must be learned to optimize use. Over time, users have learned and adapted to basic software environments that allow users to open files, download content, organize search results, and save bookmarks for later use. While once new features, these basic computing functions are now familiar and relied upon by users as a hallmark of stability in many software-based product environments. Although most information products, such as software programs, are intended to iteratively improve, they sometimes create more frustration through frequent system updates. Consequently, this underscores the importance of creating user experiences that align with users and actual use, whether through algorithmic processes, task flows, or user-centered practices in product development.

The Core Principles and Features of User Experience Design

While user experience design is informed by different schools of thought, these approaches are governed by specific principles and practices in developing and prototyping information products. Since the primary focus of user experience design emphasizes usability, it encompasses characteristics and features, which make information products more accessible, findable, responsive, and universal in use. These characteristics form the core principles which support user experience design, underscoring the importance of designing information product environments attuned to the ways users interact, think, and use those products.

One of the core features of user experience, accessible design, includes the use of elements that support consistency and stability in an information product environment regardless of a user's level of access or ability. Information design emphasizes the importance of specific sensory abilities, including hearing, sight, and touch. Therefore, developers must form a deeper understanding of both user behavior and thinking and how they affect the overall user experience with regard to access. And defining access for a particular user group can vary, depending on a wide range of contexts, limitations, and uses. For example, for users with visual-perceptual limitations, such as sight-impairment, accessibility implies the necessity of creating alternate methods of communicating visual information, while for hearing-impaired users, providing alternate methods of communication audio content is essential and necessary. Incorporating specific accessibility guidelines, such as the World Wide Web Consortium's Web Content Accessibility Guidelines (WCAG) or the US Government's Section 508 guidelines used for web design, can provide suggested techniques to create accessible alternative content. These established guidelines account for a wide range of limitations in access or ability (Web Accessibility Initiative, 2021). Accessibility has become increasingly important in user experience design, due in part to the technological evolutions in the ways in which we experience content through a wide range of devices, platforms, and products. For example, websites may incorporate information graphics, which require description for sight-impaired users. A common accessibility guideline from the WCAG requires written alternative descriptions for each graphic image, as part of the content markup, which can be interpreted by assistive software tools and essentially read aloud to users. As an additional accessibility perk that addresses technological limitations, the same alternative description can be displayed as text on the screen, which can be read by users with access problems. These alternate descriptions can also provide alternate access, of sorts, if there are problems with the visual content displaying or loading on pages. In addition, alternate methods of searching and browsing content might be provided for similar reasons, such as creating static text-based navigation menus, in place of more interactive ones, to accommodate technological and other access-based limitations. Depending on the particular audience or contextual factors present, accessible design encompasses many other techniques to support consistent and standardized user experiences. As another example, access may imply more than functional capability but be extended to creating inclusive content for particular groups of users based on cultural identity. Specific cultural factors exist in virtually all aspects of information design and content development. From the use of examples and references, direction of reading, use of colors, icons, images, shapes, and others, these factors may extend into other aspects of the user experience, such as expectation of how navigation tools are labeled, presented, positioned, or even how basic forms, used to collect information, are organized or sequenced. In this example, while the navigation menu or information form may appear to function or load properly, the perceptual or cognitive accessibility may vary for different audiences who have difficulty comprehending the meaning, presentation, or sequence in these basic functional features. Hence, their user experience may be altered by this or other information accessibility problems. This implies a broader sense of how accessibility can function within an information product, where perceived function and use may extend beyond conventional expectations. Therefore, when creating accessible features in products, it is essential to consider how to make them inherently more flexible in accommodating a range of contexts, limitations, and ultimately users.

Findability is a feature of user experience that is defined as the degree to which content is locatable and navigable, including how supportive the information environment is toward those goals (Morville, 2005).

Findability relates to both accessibility and usability of an information product environment and is often tested as a performance metric in many kinds of usability tests. Several information product characteristics, such as metadata, navigation, structure, and tagging relate to information findability. Metadata describes all aspects of content within an information product, including its behavior, properties, rules, and semantic and structural characteristics. Metadata supports many aspects of findability, including automatic reuse, improved workflow, retrieval, reporting, and status tracking within an information product environment (Rockley & Cooper, 2012). Metadata supports users through information findability, but it also supports developers by organizing and streamlining their work. Rockley and Cooper (2012) differentiate between the development methods and functions of descriptive metadata that emphasize the importance of information findability for users and content components. For developers, component metadata helps organize content into modular components for reuse and improved workflow on a project. Within information product environments, navigation and structure often share interrelated characteristics that support information findability. Categorizing content, customized sorting options, keyword tagging, and semantic markup are all features that help users comprehend both navigation and structure within an information product environment (Rockley & Cooper, 2012). When developing these features, it is important to consider how users search and browse content, which includes how they think visually and spatially. Providing multiple navigation tools that provide users choices when browsing and searching can help boost findability and performance, whether they include keyword search, site maps, tag clouds, toolbar menus, or other useful features. Also, pairing content labels, icons, images, or grouping techniques can provide useful contextual cues to assist findability. When pairing image and text, it may be helpful to use controlled vocabularies or consistent labels and terms that conform to user expectations. Also, integrating the core principles of information design can help create conceptual and semantic content within a product that supports information findability and user performance.

As another essential feature of user experience, responsive design addresses how content and environment can dynamically change, based on factors such as user preferences and system conditions. User preferences may include different layouts or styles, while system conditions can include factors such as the browser type and version, current date, operating system screen resolution, and user access level or permissions.

Responsive design elements can be integrated through the use of algorithms, content management functions, current settings or conditions, or user actions to create more engaging user experiences. Responsive design also relates to content accessibility, focusing on normalizing the user experience to better accommodate the differences in access, functionality, presentation, and technology for information product use. While users may access content using different devices and settings, responsive design attempts to account for these differences creating adaptable features that accommodate different conditions. This adaptability of both information environment (or interface) and content aims at creating a consistent user experience across the range of devices, settings, and tools used to interact with information products. For example, in a website, responsive design elements might be used to resize content boxes, images, screen sizes, or styles to accommodate different user preferences. Responsive elements can also be used to reposition content and page layouts so they function and display in consistent ways across a wide range of devices, such as laptops, phones, and tablets. Responsive designs are also reactive, which can be used to focus our attention, using new techniques or visual and spatial codes. Initially, these experiences may seem novel for some time, but eventually they may be assimilated as regular or expected patterns, or eventually discounted if they fail to serve a specific usable purpose. As a result, responsive elements may enhance access and use of an information product, in favor of creating novel ones.

Universal design, as another core feature of user experience, relates to accessible design focuses on inclusive design techniques that maximize usability in both physical and virtual environments, with minimal need for adaptation (Steinfeld & Maisel, 2012). This approach advocates a set of universal design principles, advocates for the creation of usable content for users regardless of their ability or access (US Government Section 508 Accessibility Standard; Section508.gov, 2021). Steinfeld and Maisel (2012) identify seven specific principles of universal design, which include equitable use, flexible use, simple and intuitive use, perceptible information, tolerance for error, low physical effort, and size and space appropriate for use. From an application standpoint, information products should provide equal access to content and features with enough flexibility to accommodate different limitations, modalities, and preferences with regard to use. They should also be designed simple enough that users can comprehend features and functions intuitively, with minimal effort or instruction. Content should also be presented in ways that can be easily perceived and interpreted, which relates to how users think visually and spatially in information products. Information products should also accommodate performance by providing appropriate feedback and workarounds for errors encountered in the system, and be designed to minimize effort to do so. And finally, interfaces should be responsively designed that adjust their presentation to accommodate a wide range of access parameters and devices, which may be used to interact with products. Universal design emerged from web accessibility guidelines and standards with the premise that products, places, and systems can be designed that reduce the need for special accommodations, using adaptive principles and features, which may minimize the need for potentially expensive assistive devices (Steinfeld & Maisel, 2012). Universal design principles are sometimes used to inform development practices, which ideally reduces the need for specialized accommodation or assistive tools within information product environments. When properly implemented, universal design principles and techniques can help developers provide a baseline user experience, focusing on both accessibility and usability.

Collectively, the core features and principles of user experience design overlap in both application and use, helping developers holistic, usable information products and environment designs. Ideally, user experiences should feature information products that are accessible, findable, responsive, and universal in their design and use. To be successful, these user experience features and core principles rely on successful iterative prototyping, testing, and user research to inform their continual development and improvement. User experience design attempts to balance both accessibility and usability in products to create optimal information experiences for users. When products fail to incorporate appropriate features of user experience design, they can create negative experiences that will frustrate users, who may search for more reliable product experiences.

How Design Tactics Align with Information Experience

Sometimes, information product experiences fall short of expectations and create less favorable information experiences for users. Often, the information product design, from its presentation to its interactive properties, has a significant impact on the perceived information experience. Ideally, information product design should incorporate established practices and principles that can inform specific development practices, as well as the characteristics, features, and messages of an information product. When these principles and processes are circumvented or misapplied, the information product experience can become misaligned with the intended messages and themes. Sometimes, information and user experience design results in the failure of a product to resonate with its users.

In some cases, product design practices may fail to successfully incorporate established design theories and principles, which can help developers make design choices that support users' comprehension, performance, thinking, and use of information products. Since design principles are grounded in tested theories of design, such as Gestalt theory and visual thinking, they can provide reliability in making informed design choices. Established design theories are based on basic human perceptual and cognitive processes. In turn, they support how users think and use information, including how design elements are interacted with, presented, positioned, and styled. While information products may also be informed by project-specific design conventions, such as visual identity guidelines, these elements should also align with established practices and theories. At best, designs that fail to do so may compromise the perceived value and information experience for users. Using established design principles, grounded in design theories, can help developers make optimal choices throughout information product designs.

Sometimes, the design process itself may negatively impact information product quality and, subsequently, the information experience. When designing and developing information products, an iterative and mature process approach, focusing on product innovation and sustainability will often contribute to successful product design. Mature processes rely on standardized practices, but also rely on innovation that support quality improvements over the product life cycle. Immature processes often circumvent iterative prototyping and testing or fail to improve or innovate established practices. Less mature processes may be based on ad hoc, unstructured, or unmodified processes and practices, which can result in wasted time, resources, and compromise product quality. Within an information design context, less mature processes are the result of insufficient planning, design practice, or user research. For example, an information product design scheme may include the use of a neutral color scheme, sans serif font faces, symmetrical grid layout, drop shadows for text boxes, and icon-based navigation symbols. While these specifications provide details on design features, a designer may choose to implement these features following their own collective preferences rather than established or sustained design practices, and, as a result, the product design quality may vary widely. Even sustainable practices, while often based on successful techniques, may eventually require review and updating in order to maintain and improve product designs. When similar unimproved or unstructured practices are repeated for subsequent products, there may be limitations in both innovation and quality improvement for future products. Therefore, it is important to evaluate processes to determine ways in which they can be improved and innovated over time.

Sometimes, the process may focus too much on the product or system, rather than the user in design. As a result, a system-centered design process may result in a diminished information experience when a more user-centered design one is required. Since process leads to practice, problems in the former often lead to problems in the latter. System-centered designs emphasize the product over the user, where the designer makes decisions based in internal conventions, individual preference, popular trends, or other factors independent of the user. While in some cases, a system-centered approach may be useful to maintain internal brand consistency, user research can often help innovate branded designs in ways that are more appealing to users. As an example, consider the case of a well-established shipping company, which decides they need to freshen up their brand. As a first step, they decide to poll customers about the use of new color schemes, new logos, and updated design templates to improve their website design and usability. As a sustained brand, the core design features are highly recognizable among users, and the results of the survey overwhelmingly support minimal changes in design. Despite these results, the design team decides that the brand requires a major refresh based on the latest web design trends and the support of management. As a result, the new design causes an increase in error reports and negative customer feedback on its overall usability. In a user-centered design practice, the team might have integrated user feedback to make changes to the design that aligned with the product brand. While not every bit of user research data may be useful or feasible, integrating some user feedback into the redesign effort may have led to a more favorable reception and information experience for product users.

To ensure a more favorable information experience, information product designs should be informed by established theories and principles, iterative and mature processes, and user-centered design techniques to create the optimal experience for users. As a result, the design of successful information products has a direct impact on the user's perceived information experience. The product experience will likely change over the product's life cycle, requiring an iterative sequence of user research, product redesign, and innovation in both processes and practices used. Particularly in hybrid and electronic information products, content and presentation function as interrelated components that communicate a unified information experience (Clark, 2007). And since that experience is a direct expression of an information product's value, it is especially important to align the expectations of both user and product in our design and development practices.

Implications of Tactical Design and Information Experience

The disciplines of information design and user experience design encompass a broad range of features, practices, and principles that support the development of successful information products. The tactics of design, when applied successfully, can create holistic information experiences that enhance both the content and presentation of information products. Both information design and user experience design are informed by the ways in which users perceive, interpret, and comprehend the various visual, spatial, and textual codes used in the design of both information and environment. Therefore, the tactics used in information experience design should be informed by visual and spatial thinking processes and principles, which can support the creation of more informative, interactive, and engaging information experiences for users. Design tactics must also incorporate iterative design processes that support both usability and accessibility aspects of information products. While information design focuses on features that support conceptual, consistent, contrastive, positional, and relational tactics in the presentation of information, user experience emphasizes features such as accessibility, findability, responsivity, and universal design, which promote performance. Design tactics support both presentation and interactive properties of information products. And while the specific practices used in product development may be influenced by a wide range of other product-related contexts and factors, they are essential in creating successful information product experiences.

Information design tactics are derived from Gestalt theory, focusing on core concepts and principles that emphasize conceptual, consistent, contrastive, positional, and relational aspects. While there are a wide range of practices and techniques used in information design, they are derived from the basic perceptual and cognitive acts in how users think visually and spatially about content and information environments. Conceptual design features support the use of objects and shapes that assist users in concept formation and understanding of the meaning and function of visual elements. Consistent design features include the repetition of design styles that create convention and stability in designs though style sheets and templates. Contrastive techniques incorporate figure/ground techniques, which creates emphasis and subtlety within a design environment, which supports visual perception. Positional design techniques use spatial techniques to align and group elements to demonstrate semantic function and meaning within a design, including characteristics such as disparity or similarity. Relational design strategies incorporate visual and spatial techniques to create a sense of relatedness or wholeness between elements within a design, which support the overall brand or visual identity. Collectively, the tactics of information design work symbiotically, helping designers create information product designs and tactics that support both user perception and cognition.

Information design principles inform the tactics of design, but not in mutually exclusive ways. One design tactic or technique may have multiple purposes or functions as an element of design within an information product. In fact, as illustrated through example, specific design tactics often support a singular goal, such as demonstrating similarity between objects positioned (or grouped) within in the same physical space on a page or screen. While our perception of figure and ground may be mutually exclusive, in that we can perceive the intricate details of one or the other at a time, design elements often incorporate multiple features, layers, styles, and techniques superimposed in the same space. As such, the sheer complexity of a particular design configuration (or element) may pose a perceptual challenge; however, layer by layer, we can dissect each aspect to determine its function, meaning, and message as both individual features and collective conceptual wholes. When design tactics are informed by specific design principles, or core concepts, essentially, they function as perceptual codes, which assist users in recognizing and interpreting both function and meaning within the design. Information designs can also communicate these codes in ways that support specific messages and information experiences. However, the specific information experience communicated through these design tactics will be ultimately determined by the user.

User experience design tactics incorporate iterative processes and design features that support accessibility and usability. User experience design process models include a wide range of approaches in iterative product development, which emphasize algorithmic, holistic, task-based, and user-centered practices. In turn, these processes are supported by core principles that emphasize product features that are accessible, findable, responsive, and universally designed. Accessible features include elements that can be used regardless of limitation or setting, which may include alternative content forms, multiple navigation tools, and intuitive interactive elements. Findable features ensure both content and environment can be easily browsed, located, and searched, which supports user performance of critical information tasks within a product environment. Responsive features are adaptive and accommodate a wide range of differences in the access and presentation of information products, which includes how they can be customized and personalized for users. Universal design features ensure a baseline experience for all users, regardless of their ability or access, by incorporating techniques that maximize access and use of the various features of information product environments. While user experience design emphasizes the importance of individual users (and uses) of information products, the core features of user experience design can support developers in creating optimal information product experiences.

Accessibility facilitates both function and flexibility within an information product. While an important principle of user experience, accessibility often is the first point of interaction between user and information product (or system). But access has many facets related to the overall information experience, including context, flexibility, function, meaning, and others. At its most basic level, access may imply whether or not a page loads or a particular tool functions properly. Providing alternate forms of content or functionality may be necessary to accommodate various limitations, whether they are cognitive, perceptual, physical, or even technological in nature. Physical limitations might require alternate forms of content or tools that facilitate access and use. Technological access problems are often related to overall functionality, such as whether a user can login or successfully submit a document. However, cognitive or perceptual problems might require adjustments to the organization or presentation of content, which would enable or facilitate access. In some cases, information accessibility is the problem, whereby different users have expectations for the meaning, presentation, or sequence of content within an information product. For these users, without alternate forms, their access may be impeded altogether due to their challenges with perceiving or comprehending elements within the interface or larger information environment. Access focuses specifically on whether a specific element functions or is presented to support continued performance, and not necessarily on issues related to expectation or individual preference. While problems with access typically interrupt functionality or performance, a mismatch in accommodating individual preference may simply cause frustration with users and need to be addressed as part of larger information design strategies and tactics.

Information design emphasizes the importance of presentation, while user experience underscores the need for performance. Ultimately, the tactics used in both information design and user experience design support ways in which users think and actively use information within specific working contexts. Information design involves using tactics that help establish concept, consistency, position (or placement), relation, and visual emphasis (or distinction) through the use of a wide range of presentational elements, such as color, dimensions, opacity, shading, spacing, and other stylistic features. User experience design underscores performance by focusing on access, findability, responsivity, and universality as its core concepts and principles. While the specific processes and techniques may vary between individual designers or teams, successful information product design strategies and tactics will incorporate similar core concepts and principles to inform the discrete practices and techniques used. Ideally, these practices support the integration of both content and presentation to create holistic and unified information experiences for users. As active thinkers, users learn from design environments and features to better understand information products and their various uses. Ideally, design tactics should incorporate an understanding of the user throughout the development process, beyond simple expectations and preferences, including techniques that support how they think visually and spatially in information environments. Incorporating design tactics that are supported by established core concepts and principles can assist product designers in creating more highly usable information products. And when properly executed, successful design tactics inevitably support how users experience information, rather than how they simply read or view it.

Chapter 7

Holistic Information Experiences

Information experiences are influenced by a wide range of factors that are filtered through our unique experiences, impressions, and interactions with information products and their environments. These experiences are in part a result of our perceptual and cognitive processes, as well as how we interpret brands, characteristics, messages, and themes of those products. Developers imbue information products with their own unique characteristics, features, and functions working from specific brands, product specifications, and visual identity guidelines, which represent the collective, intended information product experience to its users. Whether those products are goods or services, or physical or virtual in nature, the information experience represents a holistic combination of both developer intent and user interpretation, which can evolve over time. An information product can exist in many forms, including a wide range of print, electronic, and hybrid variants, delivered to users across a wide range of modalities and platforms, such as interactive applets, software environments, technical reports, and websites. Regardless of the communication modality, content forms, interactive media used, or product functions, they often incorporate a complex mix of features and formats, which demand more complex ways of interaction, thinking, and use.

The information experience also represents the whole effect on the user, which relies on the developer's ability to align brand and product that optimally communicate product messages. Sometimes, the intended experience may vary from the perceived experience, and it may not be possible to fully satisfy every user preference. Therefore, developers must often strive for a more generalized information experience that represents the most salient and significant characteristics of a product regardless of their intended effects. By aligning our practices and design tactics with the ways users think and use information product, developers can close the gap between intended and perceived experiences, which, although not perfect, can satisfy both product and users. Collectively, the elements of perception, cognition, branding, environment, information design, and user experience design form a holistic model of information experience that advocates content as an experience for its intended users. Whether we're using information and user experience design techniques to create styles, positional layouts, or interactive forms of content, these elements work holistically to create branded messages and meaningful experiences. This synchronization of elements represents a holistic information product experience in the user's mind, helping technical communicators create more effective information products and services for users.

Information Experiences are
Shaped by User Perception and Cognitive Processes

Our perceptions shape our initial impressions of an information experience as we focus, fixate, discern, and form concepts to form conceptual wholes from information environments (Arnheim, 1997). Our basic perceptual processes help us make sense of the world around us, whether it's a physical space, information product, object, or simulated environment. These perceptual processes are explained and informed by principles of Gestalt theory, which include how we perceive closure, continuation, figure/ ground, proximity, and similarity in information environments. In turn, these principles can serve as applied heuristics and practices that inform how developers create, design, develop, evaluate, and improve information product experiences. Our perceptual processes are also evolutionary, changing with subsequent interactions and experiences. When it comes to our perception, adaptation is the name of the game. As we encounter new information environments, our perceptual processes cycle through multiple iterations of concept formation, rejection, and refinement, evolving with continual exposure to new configurations and environments.

To illustrate, users actively focus on visual information through a hierarchy, first noticing the most distinctive, unique, and prominently placed elements in their field of vision, and, thereafter, noticing other visuals that are less distinctive or relevant (Baehr, 2002). This focus also prioritizes visual information that may help users solve specific problems or complete necessary tasks in this process of sorting and filtering information (Johnson-Sheehan & Baehr, 2001). As users explore and interact with the information environment, they evaluate objects in the foreground and background that help them determine the semantic meaning and relationships between various elements. Eventually, our initial sensory perception leads to recognition of concepts, patterns, and shapes that have both individual and collective meanings and messages. Accordingly, perception functions iteratively to help construct holistic information experiences.

While perception may be more of a reactive response, which guides our initial visual and spatial thinking, it works closely with our cognition to help us analyze and form concrete learned experiences. Our understanding of an information experience is, in part, the result of our collective visual and spatial thinking processes, which includes how we perceive and cognitively process information. Despite the differences in our experiences and interpretations, these general processes that govern perception and cognition are somewhat similar. Developers can use what they know about these basic processes to develop tactics that best align with the ways in which users interpret information, whether it is visual, spatial, or textual in nature.

Our cognition is also critical to our overall information experience, supporting how we analyze, interpret, learn, and process information environments. Arnheim (1997) suggests perception and cognition are critically interdependent processes and inextricably linked, forming the basis our visual and spatial thinking. Ultimately, together our perception and cognition form the basis of our visual and spatial thinking, which describes how users construct a holistic understanding of information experiences and their unique characteristics, features, and messages. Cognition works with our perception as an iterative exchange of sensing and making meaning. Cognition involves the tasks of comparing, contrasting, pattern matching, and semantic processing, through which we learn to use and make meaning from new and novel information environments. Collectively, these tasks function as lenses (or filters) of sorts, which support comprehension and subsequently guide our behaviors and uses of information products.

Our individual interpretations are also shaped by our prior experiences, and they may also be influenced by our preferred learning styles and motivations. For example, we may prefer (or prioritize) visual information over other sensory modalities when completing a set of instructions to assemble a cabinet, in part due to past experiences and individual preferences in completing similar tasks. However, we may prefer an auditory presentation for lectures or news stories, particularly when we are performing other tasks, such as note taking, simultaneously. Cognitive processing may also vary depending on the level of complexity in either the information environment or task-at-hand. Cognition involves multiple overlapping processes, which may vary depending on our specific needs, whether they include simple knowledge acquisition, comprehension, and application, or complex analysis, synthesis, and evaluative processes. Whether information is new or familiar, simple or complex, our cognitive processes adapt to fit our information processing needs regardless of the task, from basic rote knowledge acquisition to higher-level analyses of complex data. While these processes are adaptive, they are also evolutionary and affected by new learning experiences and sensory environments. We experience information in much the same way we learn it. Our cognitive processes are augmented by our preferred learning methods and modalities. Similarly, our comprehension depends upon both information itself and the information environment. This might include learning how content is accessed, organized, presented, or semantically coded. Once we learn new information, it becomes a learned experience, which is stored in memory as new knowledge, skills, and learned behaviors. Learned experiences function as patterns or filters, which help us adapt, learn, and perform more efficiently with other information product environments. As we learn new experiences, our stored cognitive patterns change as a result of subsequent cognitive processing. Each new information environment (or object) can be interpreted through these stored patterns or filters, helping us learn others. Developers can create information experiences that incorporate these interdependencies by using consistent, familiar, and repeated patterns and sequences that build upon prior knowledge, which can improve user performance and product experiences.

Information Environments Provide the Landscape for Information Experiences

Information environments are intended to create seamless experiences for users through interfaces that provide both the content and framework for information products. While interfaces may vary widely in their content and use, information products and technologies incorporate characteristics, features, and tools that are expected and familiar to users, to promote performance and usability in similar product environments. The interface mediates between user, content, and the system, bringing together these aspects into a seamless whole. Interfaces incorporate specific visual, spatial, and textual codes, as well as the interactive, semantic, and structural aspects of information products to create a holistic information experience. In particular, many electronic and hybrid information products share common characteristics that have foundations in hypertext theory. These characteristics include the use of associative hyperlinking, collaborative structured authoring, component content, customization, interactivity, flexible information structures, and multimodal content (Baehr & Lang, 2019).

In addition, interfaces are also influenced by changes in technological platforms and tools. The technologies used to create, develop, and present (or publish) content will also evolve and converge with other technological forms, which presents unique challenges for both developers and users in terms of access and use. For example, mobile computing devices have evolved from simple displays of text-based menus into devices with added capabilities, such as biometric data sensors, geopositional awareness, livestream video sharing, and other sophisticated tools. In turn, these changes may affect all other aspects of an information product, including its content, interactivity, and presentation, and both the developer and product must adapt. While each technological characteristic or feature may be associated with a specific era (oral, written, print, electronic), most information products include a combination of characteristics from multiple eras and forms. As an example, a portable tablet computing device includes a microphone and speakers (oral), physical writing surface and stylus (written), portable document formats that can be printed or viewed (print), and digital storage and networking capabilities (electronic). As technological products evolve, the information experience will continue to change as well. Many technological products are not isolated inventions, and, in fact, successful ones inevitably converge with others, forming new products and features that users recognize and come to expect. Consequently, developers must consider how new capabilities and features are incorporated into future product iterations to ensure they meet the evolutionary demands of these technological innovations and the changing information experiences for users.

Strategic Brands Support Successful Information Experiences

Even before users encounter them, information experiences are often carefully planned very early in the product development process, when information developers create a product's strategic brand. This creative development process may involve a wide range of brainstorming and researching techniques, such as benchmarking research, product analytics research, user profiles, and many others. Branding strategies, as planning and scoping activities, directly inform information product development, such as how specific brand messaging, content strategies, and visual identities are created and implemented. Information developers translate branded messages into visual, spatial, and textual codes, which are communicated throughout an information product and convey an intended information experience. Successful brands communicate value to users through both content and presentation, and, at the product level, users discern its characteristics, features, and messages holistically as an experience.

While a specific brand may include specific characteristics or themes, such as accurate, consistent, organized, reliable, timely, and so forth, they must be translated into messages that users can understand. Messages can include descriptions, images, keywords, slogans, taglines, and other forms, which can be enhanced by other supporting codes, whether visual, spatial, or textual in nature. These might include the use of specific colors, fonts, spatial grids, styles, or other techniques, which can convey their own messages and themes, or support existing ones. Branded messages may also be presented abstractly (such as keywords or themes that describe desired characteristics of a product), or concretely (such as specific slogans, descriptions, or taglines that provide specific detail about a product and its use). Collectively, brands function as a holistic representation of the most recognizable, salient, and sustainable features of an information experience.

While brands are primarily strategic, their supporting visual identity guidelines also function tactically. Visual identity guidelines provide the specific design conventions that communicate branded messages to users. Successful visual identity guidelines support both the product brand and the visual and spatial thinking processes of users. Brands and visual identities should also aim for a balance of both consistency and creativity in their development and implementation, which can support a product throughout its life cycle. While visual identities help establish consistent standards, they must also have enough flexibility to accommodate changes in the brands and products they support. After their initial development, strategic brands will iteratively evolve over the product life cycle and sometimes extend beyond the life of a single product. With each iteration (or version), information products must be tested and refined to ensure successful alignment with both brand and messaging. Brands must also continue to incorporate user research to better understand how to accommodate specific user expectations, preferences, and uses of products, including how these factors change newer product iterations. As an information product and its brand evolves, so does its information experience. Therefore, developers must plan for adaptability in strategic branding and continuing product development and design to ensure optimal results. In doing so, these products will be able to communicate more successful and sustainable information experiences.

Tactical Design Is the Implementation of Information Experience

Information experiences are also partially constructed by applying estab-lished design principles and practices used to create a wide range of information products. The tactics of design, which include information design and user experience design practices, incorporate a wide range of features, principles, and techniques that support the interactive properties and presentation of information products. Information design features and principles focus on the conceptual, consistent, positional, relational, and visual distinctive properties of both presentation and content. Information design principles serve as broad theoretical guidelines, which must be adapted and appropriated into specific style conventions to fit a particular information product. These conventions take the form of various style sheets and templates used for creating and designing information products. Information design principles help designers make informed choices in all aspects of the design of information products, including the structure, style, position, and presentation. User experience design focuses primary on the access and use of information product. User experience design incorporates iterative task-based processes and core principles to help maximize the usability of information products. While user experience design processes incorporate a wide range of development approaches and processes, the core features and principles focus on accessibility, findability, responsiveness, and universality of information product experiences. Often, designing user-centered products, which incorporates relevant user research holistically and iteratively throughout product design, is advocated to promote access and use. When incorporated throughout the entire product development process, user-centered design can support users throughout every aspect of product development, from planning to wireframing (or structuring) to prototyping and testing, and, finally, to product publishing.

Accordingly, information products must synchronize design elements and features, optimally, to create products that successfully influence both their perception and use (Unger & Chandler, 2012). This synchronistic characteristic suggests the importance of integrating user and product in creating optimal information experiences. Together, these practices also support how users interpret and actively use information products within various working contexts. And, as applied practices, these approaches help support the overall information experience by ensuring information products and environments support the effective presentation and interactive properties that optimize both access and use.

Challenges and Implications of Information Experiences

Despite our best intentions, information experiences are not always ideal. In some instances, our brands and messages fail to be communicated as intended, creating negative impressions for users. Sometimes, when the development team and project scope dominate user-driven concerns, this can create an imbalance or misalignment between user and product. When development teams make assumptions about user needs and preferences, based on out-of-date user research or limited anecdotal evidence, this can create a less than favorable product experience. When this misalignment occurs, such as product messages that boast excellent customer service, when paired with actual poor service, this can create credibility problems for a product's brand and overall experience. Additionally, information products often require some degree of learning the interface for optimal use, yet when these products fail to incorporate such features, they can also create negative impressions among users. Poor product information experiences often create such dissonance between intended and perceived messages. Since users ultimately interpret an information experience on their own, it is important to create development strategies and tactics that support alignment between content, environment, and user, to ensure an optimal information experience. Understanding some of the common information experience problems in greater detail can help developers avoid potential pitfalls and create more useful and well-received information products.

An over emphasis on developer or system-centered practices can contribute to poor information experiences. When developing new information products, often development teams start from scratch, with little or no user analytics to inform their initial work. For mature or updated products, sometimes developers conceive of new features, based on competitive benchmarks, new technologies, or trends, which may vary on its dependency on specific user feedback. For example, developers may create a new design concept, navigation toolbar, and other new features to include with a product update, designed to enhance the information experience. The inclusion of some features may simply be for the purposes of testing these new features in actual use. While developers may have the overall goal of improving user experience, they may lack actual user or product research data to inform or support these product changes. More often, this type of remote development thinking creates a false impression of user satisfaction, where developers assume their own experience is interchangeable with actual users. This is also indicative of a system-centered approach, where a development team thinks (and works) independently of its users and may simply decide not to conduct (or implement) relevant user research (Johnson, 1998).

As another example, think of the touch-based interfaces on mobile smartphone apps, which with a random swipe of the finger can trigger simple functions such as taking a photo or deleting a file. While developers may have thought these simple actions would aid performance of these tasks, frequent misuse or triggering of these functions unintentionally may cause frustration rather than convenience. Also, consider how some smartphones permit users to enable or disable various features or functions, such as privacy options and settings, that allow users to customize their experience. However, with subsequent smartphone operating system updates, these settings may reset to their default options or be moved to different locations within the system menus. As a result, the once simple acts of accessing and changing these features becomes a treasure hunt, of sorts, creating anxiety and frustration among users. While developers assume these changes are useful system upgrades, when not properly informed by adequate testing and use, they can create negative information experiences for users. As a result of these negative experiences or sufficient workarounds, users may migrate to other apps or products that are more intuitive, stable, and useable as a result. Understanding how to successfully incorporate features that attempt to more closely align system features with user performance can help developers avoid a mode of system-centered thinking and adopt one that supports user-centered practice.

Insufficient user-centered research and implemented design practice can create poor information experiences. User-centered design approaches that integrate user research consistently and thoroughly in product development, are widely used in many disciplines, including software engineering, game development, and other technical disciplines. Ideally, an integrative approach to user research helps information product developers understand how users interact with products, as well as their preferences, problems, uses, and workarounds. Despite the existence of such practices, developers may not always use them in such ways throughout the life cycle of product development. In some cases, a partial or sporadic integration of user research, such as during initial information product planning or even as a final round of testing prior to publication, often provides a limited perspective of how users interact with product environments. Johnson (1998) describes this as a user-friendly approach as an alternative to help improve an information product, which either has been developed using a system-centered approach or has limited resources to implement a fully developed user research and design strategy. For example, a user-friendly design approach might incorporate minor changes late in the development process, or even after product launch, to make the product more accommodating to users. Although this may mitigate some performance or usability concerns, it may only address some problems while ignoring other user concerns with this user-friendly update. This approach is often seen in mature information products that issue frequent patches or updates that address minor issues, which have been caused by new features and updates that fail to be adequately tested before publication. Ultimately, a lack of user-centered design practices is often attributed to poor planning, integration, or system-centered design approaches when developing information products. User-centered design, as a product development strategy, emphasizes the importance of incorporating user feedback and research throughout the entire development process from planning to publication to help mitigate such problems. Despite limitations in budget, resources, or time, developers can implement simple, low-fidelity solutions to gather user research to inform their development practices. Simple product analytics, surveys, and user testing can all be scaled to an appropriate scope to help ensure developers can have the data required to inform their practices and create a more user-centered information experience.

Conflicting or misaligned messages between product and user can result in negative information experiences. Successful strategic branding involves creating messages and themes that are communicated both explicitly and implicitly through various characteristics, codes, and features to information product users. When these messages are on-brand, or successful, they communicate experiences that enable users to experience the product as intended. When messages are less successful, users may attribute negative impressions based on the deficiencies encountered, regardless of intent. And when branded messages are misaligned with the user's interpretation of content and product features, the product and its reputation with users may suffer as a result. For example, developers and users may interpret branded themes, such as cutting-edge, comprehensive, fresh, and modern, in different ways, particularly when it comes to actual implementation and specific product features. For a website, developers may interpret cutting-edge and fresh as frequently redesigned and updated to reflect the latest design and technological trends. However, if these updates are too extensive and frequent, users may interpret the information experience quite differently, particularly if they encounter performance or usability issues with each update. While the intended experience is cutting-edge and fresh, it may be perceived as frustrating or unstable in its actual implementation. One solution to resolve the issue may be integrating additional user testing designed to evaluate the effectiveness of these targeted themes. This feedback may suggest other possible implementable changes to support a more aligned and ideal experience.

Strategic branding begins in the planning phase of information product development and can have a significant impact on all aspects of the information experience. Whether branded messages are direct and deliberate slogans, taglines, and descriptions, or more implicit, such as features that demonstrate desired characteristics (such as accurate, complete, intuitive, organized), these messages should be communicated in ways that align with the ways in which users think and use information product in an actual working context. Understanding how users interpret these messages, through various perceptual and cognitive processes, as well as rounds of user research and testing, can help information developers more accurately align the information experience from product to user.

Compromising basic principles of cognition and learning can negatively impact experience and performance. While not every information product has an instructional purpose, information environments are often learning experiences for users. Whether they are mastering new content, layouts, and navigation tools, or performing complex or sequenced tasks, users must learn information product interfaces for successful performance and use. When we encounter new information, contexts, and situations, we adapt and learn through our combined perceptual and cognitive abilities, which help us master new environments. In particular, adult learners have additional expectations, which must be accounted for in an information product. As users, adult learners expect relevance, involvement, control over the environment, and nontraditional or individualized learning experiences (Lee & Owens, 2004). These principles of adult learning, while essential to developing instructional content, also extend to how they learn new information environments. Their expectation of real-world application applies to both content and environment, where information must have a specific use within a working context. An information product feature must have an obvious function and purpose or it may be considered a negative factor in the overall information experience. They also prefer an active role in information environments, which may include sufficient functions and tools that support browsing, navigating, searching, sorting, and other performance-oriented behaviors. Information product environments must also accommodate their desire for flexibility in both access and use, which supports various contexts, such as changing conditions, different platforms, and individual choice. And, finally, they expect an individualized experience with information environments, which may involve the customization of features, organization, pacing, rate, or use.

Compromising one or more of these principles can result in negative impressions and information product experiences for users, particularly when they fail to see the applicability, flexibility, interactive capabilities, or usefulness of product features and functions. When users perceive product experiences as less valuable, they will likely seek out other products that better serve their unique information needs and learning preferences. Consequently, developers should conceptualize information products as learning environments, whether the product has an instructional purpose or not, that will enable them to design and implement features that support basic user cognition and learning. In doing so, developers can also align product features with user expectations, particularly when learning new or complex information environments is required.

Often, the root cause for poor product experience remains the mismatch between content, environment, and user. Since information experiences evolve with products, they can be improved with proper integration of consistent and branded messaging, user-centered design practices, applied learning principles, and other useful techniques to improve information products and experiences. Accordingly, developers must make an effort to fully understand how information experiences are perceived by users in actual working contexts to create more successful information products.

How Information Experiences Communicate Intent

Information experiences represent the holistic impressions we form about information products and environments, including the full range of branded elements, messages, product specifications, and themes that are representative of information products. Successful information experiences are communicated through implicit and explicit messages, which our perceptual and cognitive processes (and previous experiences) interpret as holistic experiences. Information experiences and products evolve together throughout various planning activities, product development tasks, marketing campaigns, and creative brainstorming, which help form the basis of the characteristics, content, and features of successful product experiences. Creating successful information experiences requires both strategy and tactics in designing content and features that support users throughout the entire product life cycle. As such, technical communicators aren't simply writing documentation for information products they create; rather, they are creating holistic information experiences with intent.

Successful information experiences are informed by visual and spatial thinking processes. Whether information products are physical, hybrid, or virtual in nature, users (and product developers) engage in similar perceptual and cognitive processes, which help them interpret the various visual, spatial, and textual codes that convey the information experience. Our perceptual and cognitive processes form the basis of our visual-spatial thinking, as we conceptualize, learn, navigate, and use information products for various purposes. The visual, spatial, and textual codes that comprise an information product act as signals for our attention, whether we are searching for content, comprehending complex concepts, performing functions, solving problems, or simply trying to make sense of the whole information environment. Visual and spatial thinking helps us perform a wide range of interpretative tasks, such as locating critical information, using navigation tools, comprehending organizational patterns and structures, and constructing conceptual meanings from the various codes and elements present. Incorporating design tactics that align with these processes can have a positive impact on the overall information experience for users, which can support the accuracy and clarity of communicated messages, reduction of user frustration, improved use of product features, helpfulness of navigation and performance tips, and many other positive aspects. Using established principles and practices that are grounded in theories of information and user experience design can also be helpful in developing information experiences that align with a user's visual and spatial thinking interpretative processes.

Successful information experiences are conceptualized as holistic experiences. While individual elements all have meaning within an information product environment, the overall information experience is the holistic sum of these collective characteristics, features, messages, and themes. Whether the environment is web-based interface, mobile device browser, software application, or print-based document, users form their impressions of an information experience collectively as conceptualized wholes. For example, when using an interactive reference guide for the first time, users process the various elements and features and classify them as useful features and functions. This may include the recognition of various content organizational patterns, navigation tools, design elements, interactive media, and other features present. Each individual classification has its own interpreted concepts and messages assigned. Individual pages, images, paragraphs, and other elements in the reference guide are also processed and interpreted at varying levels of complexity. Users also conceptualize information units, templates, and functions present, such as chapters, layers, navigation tools, sections, and so on, to aid comprehension. While we may perceive the meaning of individual elements and conceptual groups, ultimately, our information experience is constructed from our interaction with all of these elements as a singular experience. If users perceive a particular content section of information product as confusing, disorganized, and fragmented, the information experience, as a whole, is diminished. Conversely, if users perceive their interactions with an information product's keyword search feature as accurate, helpful, and intuitive, the collective information experience may be positive in nature. Despite differences in how the content section and keyword search tools are perceived, the information experience represents the sum total of these impressions, which characterize their overall interactive experiences with the product. Therefore, product development strategies should address how content and design can be aligned in ways that ensure consistent and clear messaging can be sustained throughout all aspects of an information product.

Successful information experiences are often best-fit rather than perfect-fit solutions. Creating products that align content, developer, and user is challenging enough; however, a perfect experience may be a rare occurrence, while a best-fit one may be more common with proper planning. Successful integration of strategic branding techniques and established design tactics that support user visual and spatial thinking, may help overcome many shortcomings in a product experience. In reality, developers occasionally need to make decisions and create product features that may not perfectly align with every facet of user research or expectation. Sometimes, brand or product specifications may overrule user preference to suit a particular function, limitation, or scope requirement. Some features users prefer may not be feasible under constraints set by the product or resources available. Or even quite possibly, new features may have no baseline analytics or standards in previous product iterations. From the user perspective, individual users may also appropriate information to suit their own purposes, such as devising new methods of interaction and use that more closely suits their specific needs. And while these kinds of workarounds are expected in actual product use, developers must consider how information products can be made more useful by incorporating these aspects into future product iterations. Developers should aim for a best-fit solution when integrating these factors into their product development strategies, when feasible. Best-fit solutions can also align with user-centered practices, whereby feasible user preferences can be integrated with necessary product features. Information products aligned in such a way can also help ensure longevity and improved information experiences for users. While a perfect-fit solution would ideally satisfy every individual user preference, a best-fit solution is more realistic, balancing both adaptable and feasible product designs that optimize the usefulness of information products across the widest possible user base.

Information experiences evolve over the information product life cycle for the user, as well as the product itself. Most products will iterate and change over their life cycle in many aspects of their design, features, functions, organization, and platforms. User impressions will also evolve through both continual interaction information products as their features and uses change with each product iteration. While some products have initially positive receptions among users because of their newness, novelty, or even popularity. As products evolve over time, some features are retained, some are added, some are upgraded, while others may be removed. In turn, these changes alter both the product and the information experience in different ways. Some changes may be intentional, while others may be unexpected due to unforeseen factors, such as an error, fault, or flaw in the information product. Sometimes, when one feature is upgraded or removed, this may have a cumulative effect on how other features function within the system. And in the absence of adequate testing, these faults (or bugs) may accumulate over time, causing frustration to both developer and user. For developers, it is important to be aware of these changes, particularly as users may adapt and appropriate these new features differently than expected. As users find new uses for updated or upgraded information products, they may also discover workarounds that ignore obvious faults or flaws in the system. As part of regular product development and iteration, any changes should be properly evaluated and tested to ensure the product evolutions function properly and accurately convey the desired information experience to users.

Successful information experiences align content, design, and environment in both practice and product implementation. Ultimately, the tactical work of information product developers translates strategic brands and themes into the discrete visual, spatial, and textual codes and features. Both content and design presentation share a symbiotic relationship within information products, where each supports the other (Clark, 2007). Accordingly, information development strategies should attempt to align both content and design specifications so that they communicate consistent messages throughout every aspect of a product. The alignment between content messages and product design must be continually evaluated to ensure they also are on-brand for the product. Incorporating visual and spatial thinking can support the alignment of content and presentation, specifically, principles that govern how users discern concepts and organizational patterns within information designs. Using established practices and techniques that integrate how users perceive and cognate (think visually and spatially) can aid user comprehension of information product content and features. When users can see how content and presentation relate and share semantic meaning, information products can convey more consistent and coherent information experiences. Since every permutation of user expectation and preference may not be feasible to accommodate in product design, this alignment is often critical in helping bridge the gap in understanding between new content environments and user experience.

Information Experience as Technical Communication Practice

Information experiences represent a holistic sense of an information product in the user's mind, created by a wide range of processes and practices commonly used by technical communicators. From a practical standpoint, these practices include content creation, content management, information design, interaction design, organizational design, strategic branding, user experience design, user research, and many others. Technical communicators, as well as many other content development professionals, use these collective practices to design and develop information products for a wide range of technical disciplines, users, and uses. Four commonly used practices that are critical to information experiences and covered in this book are user research, strategic branding, information design, and user experience design. While these practices have been applied and discussed through various examples, it is important to understand how the elements of information experience align with the range of knowledge, skills, and abilities within technical communication (see fig. 7.1).

Figure 7.1. The five core components of information experience. The components of information experience align with the knowledge, skills, and abilities of technical communicators and the work they do to create useful information products for a wide range of users. *Source*: Created by the author.

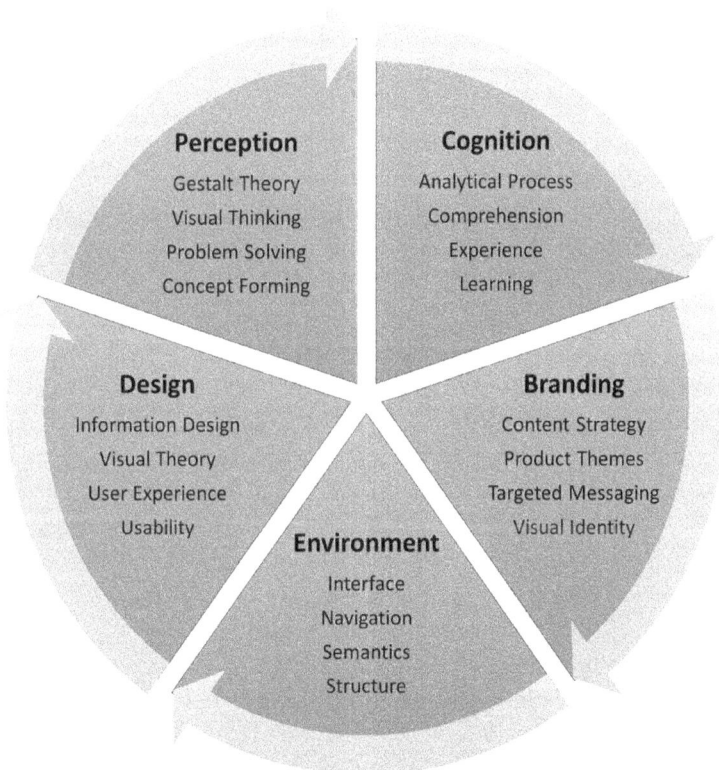

User research may typically involve collecting analytics and preferences from users and various contexts; however, this information must be interpreted with an understanding of how users perceive, comprehend, and learn from information product environments, as well. User research, as discussed in this book, focuses on understanding how users interpret and learn from information products and incorporating techniques that best support their use of these products, whether physical, hybrid, or virtual. The combined perceptual and cognitive processes that govern their interaction and use, also forms the basis of their visual and spatial thinking in information product environments. While not every information product is instructive, visual and spatial thinking supports learning basic functions of electronic information product environments that rapidly change and evolve. For example, various learning modalities can help users adapt to new information configurations, navigation tools, and methods of searching and browsing within information products. Users may focus or fixate on elements to discern their function or meaning, comparing them to learned experiences, and analyzing the conceptual relationships and semantic codes present to learn from information environments. Whether their purpose is educational, entertainment, informative, or instructive, users rely heavily on these processes to comprehend the various uses of information products. Understanding how the basic learning aptitudes and modalities are used can help developers create information products with multiple content configurations, forms, media types, and presentational formats that accommodate a wide range of users and uses. One example within instructional design products that can accommodate variety is incorporating and designing instructional architectures, which sequences content to optimize both learning and mastery. Information product structures can accommodate varied purposes and levels of difficulty by incorporating hierarchical, learner-customized structures, which allow users to control the pacing and progression of instructional content (Horton, 2011). In designing and developing information products, technical communicators are tasked with creating content for a wide range of audiences and uses that operate within a variety of contexts and constraints. To be successful in this task, user research must inform development practices, which includes understanding how to interpret user analytics through the ways in which users think visually and spatially in information product environments.

Strategic branding, which is important in product development, is also essential in creating holistic information experiences and to the work of technical communicators. As an extension of project planning, strategic branding incorporates the specific characteristics, messages, and themes that are communicated throughout all aspects of an information product. Brands can be communicated both explicitly through directed slogans or messages and implicitly, or more subtly, through suggested features and themes that convey a particular aspect of an information experience, such as accurate, comprehensive, organized, technological, and others. Often, branding strategies are developed by marketing experts; however, understanding the basics of how users, purposes, contexts, and themes are integrated into an information product is also part of the work of technical communicators. The characteristics and themes of a brand are often open to interpretation by developers in their actual implementation. One important way in which strategic branding supports information product design is through the development of visual identity guidelines that provide the positional and stylistic conventions through which various branded characteristics are communicated. For example, using *intuitive* as a branded characteristic in an instructional website might be achieved by highlighting performance tips or using animated arrows to guide users in easily completing tasks on the site. Brand characteristics often inform specific design conventions within an information product, and are combined with others to form more complex design style sheets and templates, which inform the implementation of design tactics. As part of information product development, technical communicators are critical in developing and implementing the specific tactics of a particular brand that ensure consistency and effectively communicated messages throughout an information experience.

Information design, integrates the theories and practices of creating visual, spatial, and textual codes governing the specific design tactics used in information product development. Principles of information design are founded in established design theories, including those related to perception, cognition, and visual-spatial thinking, widely used in technical communication. As one example, the principles of Gestalt theory provide foundational guidance for almost every set of principles found in design textbooks used both in academic instruction and as applied practices in industry. The primary Gestalt principles of continuation, figure/ground, similarity, and wholeness have similar corresponding terms in design texts as repetition, contrast, consistency, and completeness. The core principles of information design emphasize the conceptual, consistent, positional, relational, and visual distinctive features of design. Information design focuses on positional and stylistic properties, such as how colors, grids, images, shapes, space, and symbols communicate specific semantic and textual messages in information environments. Collectively, these design specifications represent an information product's visual identity, encompassing nearly every aspect of an information design. A product's visual identity can be communicated through established conventions, style sheets, and templates used in product design. Technical communicators rely heavily on established information design theories and principles to inform the specific tactics in creating effective designs that support successful information product experiences.

User experience design provides product developers with a wide range of processes and techniques that help create information product environments that optimize both access and use for their targeted audiences. User experience design focuses on the iterative phases of planning, development, prototyping, and testing information products that often emphasize the importance of user-centered design practices. These practices encourage the development and integration of user research analytics throughout each of the phases of user experience design to ensure products support the product functions and their users. The core features of user experience design focus on making product experiences more accessible, findable, responsive, and universal in their design, which can accommodate a wide range of users and uses. And as part of successful user experience design, information products must be deployed and tested in working contexts, so developers can select and integrate features and functions that will be most useful for task performance and use. While information products often require several iterations of prototyping and testing to achieve this, user experience design supports the use of continuous cycles of product development. These collective aspects of user experience design are also well-aligned with technical communication processes and practices commonly used in information product design and development. As part of their product development work, technical communicators routinely integrate user research analytics into user-centered design practices to create a wide range of information products, whether they are websites, interactive applications, instructional materials, electronic documents, or software-based environments. Using the collective practices of user experience design can help technical communicators create highly usable products that convey information experiences that promote both access and use for a wide range of audiences.

While these collective practices support the creation of successful information products and experiences, they also represent the critical work of technical communicators. These practices suggest the importance of aligning user, content, and environment, which communicate consistent messages and themes throughout an entire product experience. While product developers ideally want to leave users with positive, lasting impressions over the product life cycle, using both well-planned strategies and well-executed tactics can support this ultimate goal. The intentions of developers must also be aligned with the expectations of users in the finished product, which emphasize user-centered design practices. Ultimately, successful information experiences are reliant upon how well users interpret the characteristics, messages, and themes intended by product developers; however, in some cases these experiences are derived from unknown factors. Users may also appropriate products and their functions in unexpected ways that serve other purposes or as performance workarounds. Product developers can integrate iterative research, prototyping, and testing to help discover these unexpected uses and find solutions to improve overall product design.

Information experiences are the result of design thinking from both developer and user perspectives. This thinking is often visual-spatial in nature, informed by perceptual and cognitive processes and principles that are adaptive—governing our initial interaction with new information environments, but also filtered through previous experiences. Since information experiences evolve throughout various product iterations, it is important to adapt our product development practices to ensure information experiences evolve appropriately with them. Technical communicators routinely demonstrate design thinking in their work, using the combined practices of strategic branding, information design, instructional design, and user experience design. These practices also support the development of highly usable technical products that communicate technical content successfully to a wide range of users. The information experiences of these products also represent conceptual wholes that users construct from their various interactions and uses. And when content, environment, and users are properly aligned, technical communicators can create holistic and successful information experiences for their product users.

Key Takeaways from Information Experience

Within technical communication, information products encompass a wide range of content types, including applications, instructional materials, interactive media, technical reports, professional presentations, software programs, websites, and many other physical, virtual, and hybrid forms. Information experiences represent the collective information product characteristics and features, combined with the user's individual and holistic interpretations. From a user perspective, these experiences are conceived through perceptual and cognitive processes and visual-spatial thinking. From a development perspective, information experiences are the result of both strategy and tactics, encompassing applied practices in user research, strategic branding, information design, and user experience design. When content, environment, and user synchronize, products can communicate information experiences that encompass both perspectives successfully to the intended audiences. This book provides several insights on how information experiences are created, iterated, and maintained over the life cycle of information products.

Content is more than something we read; it is something we experience. Information experience theory is informed by information product specifications, while the actual information experiences are shaped by how users interpret and interact with them. The basic processes of conceptualizing, interacting, learning, and using form the basis of information experience, which suggest actions extending beyond the singular act of reading. Content is more than something to read or view; it is a holistic experience that is also based on interpretation, interaction, and use. Information product experiences create holistic and lasting impressions, many of which extend beyond the initial characteristics and features of an individual product. And while these experiences are often highly personal, they may include shared characteristics that other users may perceive and become part of the collective produce experience.

Information products and information experiences evolve together. While information products are planned with specific characteristics, features, and message in their development, the information experiences will change over time, with each iteration and continued use. Even simple changes in content, design, interaction, and organization can create unforeseen changes in how different users perceive the overall information experience. Therefore, iterative user research and testing is essential to ensure that any changes align with the intended experiences and uses of information products. User research should be integrated continually, throughout all product development phases, including creation, organization, prototyping, and testing. Content and environment are both symbiotically linked within an information product, so it is important to assess how changes in one affect the other, because, collectively, they convey a holistic information experience.

Visual and spatial thinking dictates how we experience content. Our combined perceptual and cognitive processes form the basis of our visual and spatial thinking in information environments. Perceptual processes govern the foundation of an information experience, guiding our initial impressions. Much of what we learn from an information experience begins with perception, including how we focus, fixate, discern, and conceptualize the various visual, spatial, and textual codes in information product environments. Whether we are looking at a page, physical object, screen, or simulation, the same perceptual processes govern how we interpret these information environments. However, these processes are also adaptive, allowing us to continually refine our understanding based on subsequent interactions and environmental changes. Our perceptual processes serve as the initial lens, of sorts, through which we process incoming stimuli, while our cognitive processes help us construct meaning and learn from what we perceive. From a cognitive perspective, our analytical and interpretive processes help us construct, match, recognize, and refine concepts and patterns in information environments, both simple and complex, whether abstract or concrete in nature. Whether we are performing basic knowledge and comprehension tasks or engaging in higher level cognitive feats, such as analysis and synthesis of complex patterns, our cognitive processes adapt to variations in complexity, interest, and motivation. What we find most relevant will be committed to memory as learned experiences. Our learned cognitive experiences also serve as filters through which subsequent interactions are processed, which can have cumulative effects on our information experience. Consequently, information experiences are constructed as much by users, rather than independent product features, in accordance with their visual and spatial thinking.

Information experiences represent an alignment of content, environment, and user. We experience information in much the same way we learn it—with specific preferences for its unique content and environmental characteristics, such as its design, flexibility, modality, organization, pace, relevance and semantics. Within an information product, content and environment create a holistic information experience in the user's mind. While content can be presented with various visual, spatial, and textual codes and properties, the environment has its own unique set of characteristics, which can be static, dynamic, multimodal, physical, virtual, or hybrid, depending on its unique combination of features. Information environments are often presented in hybrid combinations that include multiple characteristics, modalities, and states that accommodate how users adapt and appropriate both content and environmental characteristics and features for various uses.

Successful information experience design combines carefully planned strategies and tactics. Information experiences are the result of well-planned strategic brands and messages, both explicit and implicit. Brands take the form of characteristics, messages, and themes that are created in the planning and development processes, which must also be interpreted and implemented to successfully align content, environment, and user. Depending on how successful these communicated features are received, there may be both intended and unintended messages, which can create complexity for developers as they try to align user expectations with product specifications. For example, a user may appropriate an information product in ways unforeseen by its developers, based on an undetermined need or as a workaround for a specific problem. Branded messages are critical to the information experience, and to ensure their accurate reception by users, they must be continually evaluated and tested to ensure they result in successful information experiences for their intended users. Through various product development tactics, branded characteristics and themes become integrated features of the product, which the user interprets as part of their information experience. Information design and user experience design tactics are two relevant practices, which involve developing the various visual, spatial, and textual codes that communicate the characteristics, messages, and themes. These tactics also emphasize developing features of an information environment that make information more accessible, findable, responsive, and universal for specific uses and users. Collectively, the strategies and tactics of information design establish a product's baseline information experience, which ideally aligns with the ways in which users will appropriate, interpret, and experience them.

Technical communicators create more than content; they create holistic information experiences. While many disciplines may integrate individual practices and principles of developing information products and environments, technical communication brings together the core competencies of developing all aspects of an information experience. These competencies include the critical tasks in information experience design, including user research, strategic branding, environment design, and the tactics of information design and user experience design. Collectively, these practices enact design thinking that enables us to create successful technical products for a wide range of users, purposes, and contexts. The products we create are more than useful content products created for various technical industries and purposes; rather, they are holistic information experiences, which are accessible, engaging, intuitive, responsive, and useful to the ways in which best serve our intended users.

References

Airey, D. (2019). *Identity designed: The definitive guide to visual branding*. Rockport Publishers.

Andersen, R., & Hackos, J. (2021). Practicing technical communicators' experiences with and perspectives on academic publishing. *Technical Communication*, *68*(3), 29–55.

APM Group International. (2024). *Certifications and solutions*. https://apmg-international.com/our-services/certifications

Arnheim, R. (1997). *Visual thinking*. University of California Press.

Baehr, C. (2024). *The agile communicator: Principles and practices in technical communication* (4th ed.). Kendall Hunt.

Baehr, C. (2010). Thinking visually: Heuristics for website analysis and design. In S. Josephson, S. B. Barnes, & M. Lipton (Eds.), *Visualizing the web: Evaluating online design from a visual communication perspective* (pp. 85–98). Peter Lang.

Baehr, C. (2007). *Web development: A visual-spatial approach*. Prentice-Hall.

Baehr, C., & Lang, S. M. (2012). Hypertext theory: Rethinking and reformulating what we know, Web 2.0. *Journal of Technical writing and Communication*, *42*(1), 39–56.

Baehr, C., & Lang, S. (2019). Hypertext theory: Theoretical foundations for technical communication in the 21st century. *Technical Communication*, *66*(1), 93–104.

Baehr, C., & Schaller, B. (2010). *Writing for the Internet: A guide to real communication in virtual space*. ABC-CLIO.

Barry, A. M. (1997). *Visual intelligence: Perception, image, and manipulation in visual communication*. State University of New York Press.

Berners-Lee, T. (1999). Realising the full potential of the web. *Technical Communication*, *46*(1), 79–83.

Black, A., Luna, P., Lund, O., & Walker, S. (2017). *Information design: Research and practice*. Routledge.

Bloom, B. S., Engelhart, M. D., Furst, E. J., Hill, W. H., & Kratwohl, D. R. (1956). *Taxonomy of educational objectives*. Longmans, Green.

Bloomstein, M. (2021). *Trustworthy: How the smartest brands beat cynicism and bridge the trust gap*. Page Two Books.

Boettger, R. K., & Friess, E. (2020). Content and authorship patterns in technical communication journals (1996–2017): A quantitative content analysis. *Technical Communication, 67*(3), 4–24.

Bolter, J. D. (2001). *Writing space: Computers, hypertext, and the remediation of print*. Routledge.

Bolter, J. D., & Grusin, R. (2003). *Remediation: Understanding new media*. MIT Press.

Bush, V. (1945). As we may think. *Atlantic Monthly, 176*(1), 101–108.

Carradini, S. (2020). A comparison of research topics associated with technical communication, business communication, and professional communication, 1963–2017. *IEEE Transactions on Professional Communication, 63*(2), 118–138.

Clark, D. (2007). Content management and the separation of presentation and content. *Technical Communication Quarterly, 17*(1), 35–60.

Crawford, C. (2003). *The art of interactive design*. No Starch Press.

Fleming, N. D., & Mills, C. (1992). Not another inventory, rather a catalyst for reflection. *To Improve the Academy, 11*(1), 137–155.

Gardner, H. E. (2008). *Multiple intelligences: New horizons in theory and practice*. Basic Books.

Garrett, J. J. (2010). *The elements of user experience: User-centered design for the web and beyond* (2nd ed.). Pearson Education.

Gordon, W. (1989). *Learning and memory*. Brooks/Cole.

Governor, J., Hinchcliffe, D., & Nickull, D. (2009). *Web 2.0 architectures*. O'Reilly Media.

Hackos, J. (2007). *Information development: Managing your documentation projects, portfolio, and people*. Wiley.

Hart, H., & Baehr, C. (2013). Sustainable practices for developing a body of knowledge. *Technical communication, 60*(4), 259–266.

Heim, M. (1999). *Electric language: A philosophical study of word processing*. Yale University Press.

Heim, M. (2000). *Virtual realism*. Oxford University Press.

Horton, W. (2011). *E-learning by design*. Wiley.

IDEO. (2024, June 13). What is design thinking? [Blog post]. https://www.ideou.com/blogs/inspiration/what-is-design-thinking

Jenkins, H. (2006). *Convergence culture*. New York University Press.

Johnson, R. R. (1998). *User-centered technology: A rhetorical theory for computers and other mundane artifacts*. State University of New York Press.

Johnson, S. (1997). *Interface culture: How new technology transforms the way we create and communicate*. Harper.

Johnson-Sheehan, R. (2024). *Technical communication today* (7th ed.). Pearson/Longman.

Johnson-Sheehan, R., & Baehr, C. (2001). Visual-spatial thinking in hypertexts. *Technical communication*, 48(1), 22–30.

Kimball, M. A. (2017). "Tactical technical communication." *Technical Communication Quarterly*, 26(1), 1–7.

Koffka, K. (1935). *Principles of Gestalt psychology*. Harcourt.

Kohler, W. (1947). *Gestalt psychology*. Liveright.

Kolb, D. A. (1976). *The learning style inventory: Technical manual*. McBer.

Kolb, D. A. (1984). *Experiential learning: Experience as the source of learning and development* (Vol. 1). Prentice-Hall.

Kostelnick, C. (1996). Supra-textual design: The visual rhetoric of whole documents. *Technical communication quarterly*, 5(1), 9–33.

Kostelnick, C., & Roberts, D. D. (1998). *Designing visual language: Strategies for professional communicators*. Longman.

Landow, G. P. (2006). *Hypertext 3.0: Critical theory and new media in an era of globalization*. Johns Hopkins University Press.

Lang, S., & Baehr, C. (2023). Hypertext, hyperlinks, and the World Wide Web. In O. Kruse, C. Rapp, C. Anson, K. Benetos, E. Cotos, A. Devitt, & A. Shibani (Eds.), *Digital writing technologies in higher education: Theory, research, and practice* (pp. 51–61). Springer.

Lanham, R. A. (2010). *The electronic word: Democracy, technology, and the arts*. University of Chicago Press.

Lee, W. W., & Owens, D. L. (2004). *Multimedia-based instructional design: Computer-based training, web-based training, distance broadcast training, performance-based solutions*. Wiley.

Luft, J., & Ingham, H. (1961). The Johari Window: A graphic model of awareness in interpersonal relations. *Human Relations Training News*, 5(9), 6–7.

McLeod, S. A. (2017, October 24). Kolb's learning styles and experiential learning cycle. *Simply Psychology*.

McLuhan, M. (2017). *The medium is the massage*. Routledge.

Merrilees, G. (2021, October 21). How to create a powerful brand identity. *Studio 1 Design's Blog*. https://studio1design.com/how-to-create-a-powerful-brand-identity/

Moore, L. E., & Earnshaw, Y. (2020). How to better prepare technical communication students in an evolving field: Perspectives from academic program directors, practitioners, and recent graduates. *Technical Communication*, 67(1), 63–82.

Morville, P. (2005). *Ambient findability: What we find changes who we become*. O'Reilly Media.

Morville, P., & Rosenfeld, L. (2006). *Information architecture for the World Wide Web: Designing large-scale websites* (3rd ed.). O'Reilly Media.

Nelson, T. H. (1992). *Literary machines*. Mindful Press.

Newmark, J., & Bartolotta, J. (2021). Creating the "through-line" by engaging industry certification standards in SLO redesign for a core curriculum technical writing course. In M. J. Klien (Ed.), *Effective teaching of technical communication: Theory, practice, and application* (pp. 147–165). University Press of Colorado.

Norman, D. A. (2013). *The design of everyday things*. MIT Press.

Ong, W. J. (1983). *Orality and literacy*. Routledge.

Reed, S. (2022). *Cognition theories and applications* (10th ed.). Sage.

Richey, R., Klein, J., & Tracey, M. (2011). *The instructional design knowledge base: Theory, research, and practice*. Routledge.

Rockley, A., & Cooper, C. (2012). *Managing enterprise content: A unified content strategy*. New Riders.

Rude, C. D. (2009). Mapping the research questions in technical communication. *Journal of Business and Technical Communication, 23*(2), 174–215.

Section508.gov. (2021). *Universal Design and Accessibility*.

Smith, G. (2008). *Tagging: People-powered metadata for the social web*. New Riders.

Smith, P. L., & Ragan, T. J. (2005). *Instructional design* (3rd ed.). Wiley.

Society for Technical Communication. (2024). *Certified professional technical communicator study guide*. https://www.stc.org/wp-content/uploads/2020/04/cptcstudyguide-Foundation.pdf

Steinfeld, E., & Maisel, J. (2012). *Universal design: Creating inclusive environments*. Wiley.

Technical Communication Body of Knowledge (TCBOK). (n.d.). Society for Technical Communication. http://www.tcbok.org

Unger, R., & Chandler, C. (2012). *UX-design: A practical guide to designing interaction experiences*. 2nd ed. Symbol-Plus.

VARK Learn Limited. (n.d.). *VARK Modalities: What Do Visual, Aural, Read/write, and Kinesthetic*. https://vark-learn.com/introduction-to-vark/the-vark-modalities/

Web Accessibility Initiative. (2021). *Web Content Accessibility Guidelines (WCAG) overview*. Retrieved July 7, 2021, from https://www.w3.org/WAI/standards-guidelines/wcag/

Williams, R. (2015). *The non-designer's design book: Design and typographic principles for the visual novice*. Pearson Education.

Index

www.ingramcontent.com/pod-product-compliance
Lightning Source LLC
Chambersburg PA
CBHW031127270326
41929CB00011B/1530